Last Oasis

OTHER NORTON/WORLDWATCH BOOKS

Lester R. Brown et al.

State of the World 1984	*State of the World 1994*
State of the World 1985	*State of the World 1995*
State of the World 1986	*State of the World 1996*
State of the World 1987	*State of the World 1997*
State of the World 1988	*Vital Signs 1992*
State of the World 1989	*Vital Signs 1993*
State of the World 1990	*Vital Signs 1994*
State of the World 1991	*Vital Signs 1995*
State of the World 1992	*Vital Signs 1996*
State of the World 1993	*Vital Signs 1997*

ENVIRONMENTAL ALERT SERIES

Lester R. Brown et al.
Saving the Planet

Alan Thein Durning
How Much is Enough?

Sandra Postel
Last Oasis

Lester R. Brown
Hal Kane
Full House

Christopher Flavin
Nicholas Lessen
Power Surge

Lester R. Brown
Who Will Feed China?

Lester R. Brown
Tough Choices

Michael Renner
Fighting for Survival

LAST OASIS

Facing Water Scarcity

Sandra Postel
With a New Introduction

The Worldwatch Environmental Alert Series
Linda Starke, Series Editor

W · W · NORTON & COMPANY
NEW YORK LONDON

TD
345
.P67
1997

First Edition

The text of this book is composed in Plantin
with the display set in Zapf Book Medium.
Composition and manufacturing by the Haddon Craftsmen, Inc.

ISBN 0-393-31744-7 (pbk)

W.W. Norton & Company, Inc., 500 Fifth Avenue, New York, N.Y. 10110
W.W. Norton & Company Ltd., 10 Coptic Street, London WC1A 1PU

1 2 3 4 5 6 7 8 9 0

 This book is printed on recycled paper.

Contents

Introduction to the New Edition

In early 1992, I travelled through the Middle East doing research in preparation for writing *Last Oasis*. The impressions I took away, and that informed my writing, remain strong. There was the green sliver of fertile earth alongside the Nile River, supporting tens of millions of Egyptians; the river Jordan's paltry flow, the source of so much tension in the region; Bedouin farmers on the West Bank who were growing tomatoes with drip irrigation methods; an Israeli kibbutz in the western Galilee treating and using wastewater from a neighboring Arab village to irrigate fields of cotton. They were images of tension and vulnerability, but also of cooperation and hope.

In writing *Last Oasis*, I was time and again aware of these diverging paths leading to very different—but possible—futures. This Introduction to the New Edition gives me the chance to update some of the trends and issues discussed in the book and to reflect on how far we have moved along each path during the last five years. The occasion for this new edition is the release of a documentary film for U.S. public television based on *Last Oasis*. It is the final episode of a four-part series entitled "Cadillac Desert," which we hope will be the springboard for one of the largest public engagement campaigns on water issues ever undertaken.

The first three episodes of "Cadillac Desert" chronicle the story of water development in the western United States and the power and politics behind the damming and diverting of western rivers. The fourth episode, called "Last Oasis" and created by Boston-based producer Linda Harrar, brings to life the key themes of this book: how the dam-building frenzy moved abroad, how water scarcity threatens human health and regional peace, and how the ecosystems on which all life depends urgently need protection as human demands for water spiral upward. The film drives home how the untapped potential of conservation is, increasingly, our "last oasis."

My current thinking about water issues has been influenced by recent visits to two of the most degraded waterscapes on earth. The first was the disaster zone of the Aral Sea in Central Asia. Reading reams of facts and figures on the devastation in this area had not prepared me for the emotional impact of standing on a bluff outside the former port town of Muynak and seeing only a dry, degraded, salted-out seabed with ships rotting in the sand. The sea was some 30 kilometers away, having lost half its area and three fourths of its volume.

A simple judgment by Moscow officials that the rivers of the Aral Sea basin were worth more growing cotton in the desert than sustaining the planet's fourth largest lake had produced an environmental disaster rivaling Chernobyl. The people of the region, who had no say in the decision, now suffer the consequences—destroyed fisheries, lost jobs, toxic surroundings, and high rates of respiratory, infectious, and parasitic diseases that have caused their life expectancy to plummet.

Tempting as it was, I could not pass off the Aral Sea tragedy as an unfortunate legacy of communist central planning, because a little over a year later I visited a place and a people in North America that were victims of the same sort of disregard for nature. Linda Harrar and I travelled to the Colorado River delta in northern Mexico in May 1996 to see if there was a story there to be filmed. The answer was obvious. Here was the place the great American naturalist Aldo Leopold, after canoeing there in 1922, had called a "milk and honey wilderness" full of "a hundred green lagoons."[1] Today, the delta is a desiccated place of mud-cracked earth, salt flats, and murky pools.

Virtually the entire Colorado's flow is captured and siphoned off upstream to fill swimming pools in Los Angeles, generate electricity for Las Vegas, and irrigate crops in the deserts of Arizona, California, and the Mexicali Valley. Most of the abundant wildlife Aldo Leopold saw in the delta are now gone. And the Cocopa—the "people of the river" who have fished and farmed in the region for many centuries—are now a culture at risk of extinction. "Without water," a Cocopa elder told us, "everything is lacking." And yet, holding fast to his tribe's historical way of life, he added, "I hope to see the Colorado River rise again."

The Aral Sea and the Colorado delta reveal the endpoints of a path that reflects the view that the human economy is separate from the larger ecosystem around it rather than a dependent part of it. This Introduction and the remaining chapters of *Last Oasis* cite many other examples that, while less extreme, indicate that much of the world is on this same path. I remain convinced that the final chapter, "A Water Ethic," is the most important

[1] Aldo Leopold, *A Sand County Almanac* (New York: Oxford University Press, 1949).

one, for without a deeper appreciation for water's fundamental role as the basis of life, we will keep whittling away its life-sustaining ecological functions. Drip irrigation, low-flush toilets, and other efficiency measures are critically important because they enable us to do more with less. But if we use the water saved through efficiency merely to fill more swimming pools, to irrigate more golf courses, and to support millions more meat-rich diets, we will get no closer to a sustainable world; we will simply allow unsustainable levels of consumption and population growth to persist a bit longer.

Sustainably providing the water needed for an expected population of 8 billion to have adequate and safe drinking water, food for a nutritious diet, and sufficient material goods poses enormous challenges. Indeed, given that about a fifth of our current population of 5.8 billion lack these, the task is downright daunting.[2] The only way to lessen conflict and social unrest in a world where a resource as basic as water becomes scarce is to ensure that enough is provided for all before some people get more than enough. Ultimately, a water ethic is about sharing—both with nature and with each other.

My hope is that the documentary film and this new edition of *Last Oasis* will hasten the evolution of such an ethic, and will help foster the civic responsibility and political action to bring that ethic to life.

Water Limits Draw Closer

Water is renewable, but it is also finite. As population and consumption grow in any given location, water demand approaches the natural limits of the water supply. If it hits those limits, signs of trouble emerge—such

---

[2] United Nations Commission on Sustainable Development (UN-CSD), *Comprehensive Assessment of the Freshwater Resources of the World* (New York: United Nations, 1997); U.N. Food and Agriculture Organization (FAO), *Food for All* (Rome: 1996); U.N. Development Programme (UNDP), *Human Development Report 1996* (New York: Oxford University Press, 1996).

as falling water tables, dried-up rivers, and shrinking lakes and wetlands. As documented in *Last Oasis*, these physical symptoms of scarcity are now evident in many parts of the world. But is the world as a whole likely to bump into water limits anytime soon?

I decided to explore this question in greater depth in 1995, along with biologists Gretchen Daily and Paul Ehrlich of Stanford University. We set out to answer two basic questions. First, we wanted to know how much of the earth's vast amount of water is fresh, renewable, and accessible. Most statistics on water availability do not differentiate between water than can be tapped for irrigation, industries, and households and water that cannot—either because it is too remote or because it is floodwater that dams have not captured. Second, we wanted to know how much of this accessible supply humans already use—either by withdrawing it from rivers, lakes, and aquifers for human activities, or by relying on it within its natural channel to dilute pollution, sustain fisheries, or satisfy other "instream" needs. We hoped this would give us some perspective on the sustainability of global water trends.

The results surprised us. We found that only 31 percent of the total amount of water that annually flows toward the sea is actually accessible to humans. We also found that the human economy is already using 35 percent of this accessible runoff to supply irrigation, industries, and households and at least an additional 19 percent of it to meet instream needs. So humanity is now appropriating more than half of accessible runoff.[3]

Initially, this may not sound so terrible: we can choose to view the glass as half full rather than half empty. What really matters, however, is how many people the remaining half must satisfy, how fast new water demands will rise, how much additional water technology can make

[3] Sandra L. Postel, Gretchen C. Daily, and Paul R. Ehrlich, "Human Appropriation of Renewable Fresh Water," *Science,* February 9, 1996.

available at an affordable cost, and what is happening to the health of the environment and other species because humans so dominate this basic life-support resource. By these measures, the picture is hardly sanguine.

Demand for water, like that for many resources, has been growing exponentially. If world water demand continues to grow faster than population (as it has since 1950), say at 2 percent a year, it will double in 35 years. Given that water demand tripled between 1950 and 1990 (see Chapter 3), a doubling over the next 35 years seems quite possible. Meanwhile, it is getting harder to expand the accessible supply. The construction of dams—which capture runoff and thus bring more water under human control—has slowed markedly over the last couple of decades as the public, governments, and financial backers have begun to pay more attention to the high economic, social, and environmental costs of dams. Whereas nearly 1,000 large dams began operation each year from the fifties through the mid-seventies, the number dropped to about 260 during the early nineties.[4] Even if conditions become more favorable to dam construction, it seems unlikely that new reservoirs built over the next 30–35 years will increase accessible runoff by more than 10 percent.

For several decades, the desalting of seawater has held out the promise of limitless supplies of fresh water, yet this source remains a minor contributor to the global supply picture. Global desalination capacity is now half again as large as the 1990 capacity reported in Chapter 3, "Engineering's Promise," but it still accounts for less than 0.2 percent of world water use.[5] After three decades of intensive research and development, removing salt from seawater remains energy-intensive and

---

[4] Patrick McCully, *Silenced Rivers: The Ecology and Politics of Large Dams* (London: Zed Books, 1996).

[5] Desalination capacity from Pat Burke, Secretary General, International Desalination Association, Topsfield, Mass., private communication, August 1, 1996.

expensive. The world's water situation would no doubt be very different today if as much money and effort had gone into improving water efficiency as has gone into desalination. Although desalination will continue to provide drinking water to water-poor countries that can afford it, its contribution to the global supply picture is likely to remain small for the foreseeable future.

It may be hard to believe that water limits are drawing closer, for we hold in our minds an image of earth as a strikingly blue planet—a world of water spinning in space—thanks in part to the beautiful photographs taken by astronauts. But this picture creates a false sense of security, because we can tap only a tiny fraction of this water wealth, and that small share must sustain not only our growing population but millions of other species. Limits to our use of fresh water for the things we need— clean water to drink, food to eat, material sufficiency, and a healthy environment—are on the horizon. As documented in the second half of *Last Oasis*, the best way to avoid costly shortages lies in reducing our water demand through conservation, efficiency, and better management. Meanwhile, as we struggle through the shift from a world of plenty to one of scarcity, we face some serious challenges—to food security, ecosystem health, and social and political stability.

Water for Food

Growing the food needed for a nutritious but low-meat diet for one person for a year takes about 1,100 cubic meters (nearly 291,000 gallons) of water.[6] In humid climates, rainfall delivers virtually all this needed moisture to the soil. But in less humid regions and in those with distinct wet and dry seasons, a portion of the needed

---

[6] Based on Wulf Klohn and Hans Wolter, "Perspectives of Food Security and Water Development," unpublished paper.

moisture must be supplied by irrigation. If 40 percent of the water required to produce an acceptable diet for the 2.4 billion people expected to be added to the planet over the next 30 years has to come from irrigation, agricultural water supplies would have to expand by more than 1,750 cubic kilometers per year—equivalent to roughly 20 Nile Rivers, or to 97 Colorado Rivers.

It is not at all clear where that water could come from on a sustainable basis. Over the last five years, the constraints discussed in Chapter 4, "Bread and Water," have become more pronounced. Per capita irrigated area, for example, has continued to decline, having fallen 7 percent from its 1979 peak as population growth outpaced the spread of irrigation.[7] As much as 2 million hectares of irrigated land—an area a bit larger than Kuwait and equal to nearly 1 percent of world irrigated area—comes out of production each year because of waterlogging and salinization of soils.[8] With additional irrigated land being lost to urbanization, David Seckler, Director General of the International Irrigation Management Institute, concludes that "the net growth of irrigated area in the world has probably become *negative*."[9]

Groundwater overpumping—another unsustainable practice—continues to plague future food production in some of the world's most important crop-producing regions, including the U.S. High Plains, California's Central Valley, the north China plain, and portions of India. In India's Punjab, for example, where a highly productive rice-wheat cropping pattern has turned the

---

[7] Gary Gardner, "Irrigated Area Dips Slightly," in Lester R. Brown, Christopher Flavin, and Hal Kane, *Vital Signs 1996* (New York: W.W. Norton & Company, 1996).

[8] Dina L. Umali, *Irrigation-Induced Salinity* (Washington, D.C.: World Bank, 1993).

[9] David Seckler, *The New Era of Water Resources Management: From 'Dry' to 'Wet' Water Savings* (Washington, D.C.: Consultative Group on International Agricultural Research, 1996).

region into the nation's breadbasket, water tables are falling 20 centimeters annually over two thirds of the state. According to researchers at Punjab Agricultural University, "questions are now being asked as to what extent rice cultivation should be permitted in the irrigated Indo-Gangetic Plains, and how to sustain the productivity of the region without losing the battle on the water front."[10]

Agriculture is also losing some of its existing water supplies to cities as population growth and urbanization push up urban water demands. Worldwide, the number of urban dwellers is expected to double, to 5 billion, by 2025.[11] Pressure to shift water from farms to cities is thus bound to intensify—as is already happening in China, the western United States, and other water-short areas.

Casting a disturbing shadow over all these trends is the fact that limited water supplies combined with population growth appear to be eliminating the option of food self-sufficiency in more and more countries. At runoff levels below 1,700 cubic meters per person, food self-sufficiency is often highly problematic, if not impossible. Of the 28 countries in Africa and the Middle East that are at or below this benchmark, 19 already import at least 20 percent of their grain. As populations grow, more countries will join the water-stressed list, and those already on the list will acquire more people. Thus, dependence on grain imports is likely to deepen and spread. By 2025, Africa and the Middle East alone will have more than 1.3 billion people living in water-stressed countries, up from 380 million today. In Asia, India will join the list by 2025, and China will only nar-

---

[10] International Rice Research Institute, *Water: A Looming Crisis* (Manila: 1995).
[11] Gershon Feder and Andrew Keck, "Increasing Competition for Land and Water Resources: A Global Perspective," paper prepared for World Bank, Washington, D.C., March 1995.

rowly miss doing so by then.[12]

All told, as many as 3.6 billion people could be living in countries where water supplies are too limited for food self-sufficiency. This raises some important questions. How much grain will these countries collectively need to import? Who will supply that grain, and at what price? Will importing countries, particularly those in Africa, be able to pay that price? With more than 1 billion people living in acute poverty and some 840 million people lacking sufficient food even in today's world of 5.8 billion, avoiding these issues could have severe repercussions in a world of 8 billion.[13]

An all-out effort to raise the water productivity of the global crop base—both irrigated and rain-fed—is urgently needed. This will take more widespread use of the conservation technologies and methods described in Chapters 8 and 9—from drip irrigation to soil terracing—which will only come about with changes in policies and incentives, as discussed in Chapter 13. Urban wastewater could become an important supply for agriculture (see Chapter 10), since it will be one of the few sources that is both increasing and reliable. Treated wastewater now accounts for 30 percent of Israel's agricultural water supply, for example, and this figure is expected to rise to 80 percent by 2025.[14] Better matching of crops to varying qualities of water and the breeding of new varieties that are more salt-tolerant, water-efficient, or drought-resistant will be critical to sustaining crop production in the new era of water con-

---

[12] Africa runoff figures from FAO, *Irrigation in Africa in Figures* (Rome: 1995); other runoff figures from World Resources Institute, *World Resources 1994-95* (New York: Oxford University Press, 1994); population figures from Population Reference Bureau, *1995 World Population Data Sheet* (Washington, D.C.: 1995); net grain imports from U.S. Department of Agriculture, Economic Research Service, "Production, Supply, and Distribution," electronic database, Washington, D.C., updated February 1996.

[13] UNDP, *Human Development Report 1996;* FAO, *Food for All.*

[14] Hillel Shuval, presentation at the Stockholm Environment Institute/United Nations Workshop on Freshwater Resources, New York, May 18-19, 1996.

straints—especially if the wild card of climate change brings on more drought. The development of new grain varieties will not happen overnight, however, or without stepped-up public-sector commitment.

Many countries still do not have a clear picture of water-food linkages, and thus are not taking the actions needed to secure their agricultural bases—whether this be limiting the construction of golf courses, as Vietnam and China have done; rehabilitating salinized soils; improving the efficiency of irrigation systems; or moving to secure equal land and water rights for women, the lack of which is thwarting human development and agricultural productivity in many parts of the world.

Repairing and Protecting Aquatic Ecosystems

Healthy, well-functioning rivers and aquatic ecosystems are as fundamental to the workings of the natural world as arteries, veins, and the heart are to a human body. Their decline represents a crippling of the planet's circulatory system and a crumbling of its ecological foundation.

Freshwater ecosystems—which include rivers, floodplains, lakes, swamps, wetlands, and deltas—perform a host of vital functions. Rivers, for instance, deliver nutrients to the seas and so nourish marine food webs. They sustain fisheries, dilute our waste products, provide convenient shipping channels, create habitat for a rich diversity of aquatic life, maintain soil fertility, and offer us some of the most inspirational natural beauty on the planet. These functions are easy to take for granted because they are rarely priced by the market, and they require virtually no investment on our part. Their value to us, however, is enormous. Aquatic ecologist Stephen Carpenter of the University of Wisconsin and I have estimated that the total global value of the benefits and services provided by freshwater systems, while impossible to quantify accurately, almost certainly amounts to several trillion dollars annually. We estimated that pollution dilution and transportation services alone are worth at least $700 billion a year.[15]

These ecosystem services are now deteriorating rapidly. Numerous human activities—from the construction of dams, dikes, and levees to uncontrolled pollution, excessive river diversions, and the draining of wetlands—are destroying ecological functions before they have been properly valued, or sometimes even identified. This is the key message of Chapter 5; since it was written, more evidence has come in.

In the United States, the most comprehensive assessment to date of the status of native plant and animal species was released in 1996 by a private group, The Nature Conservancy. Its striking finding is "the dire condition of those species that depend on aquatic systems for all or part of their life cycle." The study found that 67 percent of freshwater mussels are at risk, along with 65 percent of the crayfish, 38 percent of the amphibians, and 37 percent of freshwater fish. Out of 13 groups, the 4 with the greatest share of species at risk all depend on aquatic ecosystems. The leading cause of these species' imperilment, the study concluded, is the destruction and degradation of their habitats.[16]

Similar assessments have not yet been done for most countries. Habitat destruction, however, is without question a pervasive and worsening stress on most aquatic systems. Around the world, for example, the number of large dams—one of the primary destroyers of river habitat—has climbed from just over 5,000 in 1950 to roughly 40,000 today.[17] More than 85 percent of the large dams now standing have been built during the last 35

---

[15] Sandra Postel and Stephen Carpenter, "Freshwater Ecosystem Services," in Gretchen C. Daily, ed., *Nature's Services: Societal Dependence on Natural Ecosystems* (Washington, D.C.: Island Press, 1997).

[16] The Nature Conservancy, *Priorities for Conservation: 1996 Annual Report Card for U.S. Plant and Animal Species* (Arlington, Va.: 1996).

[17] McCully, *Silenced Rivers*.

years, a massive change in the global aquatic environment in a short period of time.

In doing research for an article in 1995, I was struck by the number of major rivers in the world that no longer reach the sea for large portions of the year. The Nile in Egypt, the Ganges in South Asia, the Amu Dar'ya and Syr Dar'ya in Central Asia, the Yellow River in China, and the Colorado River in North America are among the major rivers that are so dammed, diverted, or overtapped that little or no fresh water reaches its final destination for significant stretches of time.[18] China's Yellow River has gone dry in its lower reaches for an average of 70 days a year in each of the last 10 years; in 1995, it was dry for 122 days.

India's heavy diversions of the Ganges have left little water for Bangladesh during the dry season, much less enough to reach the river's natural outlet in the Bay of Bengal. The result has been the rapid advance of a salt front across the Ganges delta, damaging valuable mangroves and fish habitat. And the heavy diversions of the Amu Dar'ya and Syr Dar'ya rivers in the Aral Sea basin and of the Colorado River have produced disturbing evidence of social and environmental destruction. Toward the end of the "Last Oasis" documentary, Alejandro Robles, executive director of Conservation International's Mexico program, reflects on the sad state of the Colorado delta: "We should not repeat what we did with the Colorado River, because those things do not reverse."

The silver lining is that efforts to restore and protect aquatic ecosystems are gaining force, thanks in large part to the dedicated work of nongovernmental organizations (NGOs), citizen activists, scientists, and others who have given political voice to ecosystems and the people most dependent on them. Although many of the responses are too little or too late, there is hope that at least some of

---

[18] Sandra Postel, "Where Have All the Rivers Gone?" *World Watch*, May-June 1995.

the damage can be repaired, and that hard-learned lessons might prevent irreversible mistakes from being repeated.

Restoring the Aral Sea and surrounding delta ecosystems to anything like they were 50 years ago, for example, is impossible. Even stabilizing the sea at its present level appears overly optimistic, since this would require more river water than likely could be freed up.[19] But some modest repair, such as constructing wetlands in the river deltas, may be feasible. A multidonor program coordinated by the World Bank includes a proposal to create a small freshwater lake in the northern part of the Aral by diking it off from the rest of the salty sea and ensuring that some river water flows into it. Such efforts cannot compensate for what has been lost, but they might allow some native vegetation and fisheries to return, generate local jobs, and permit some ecological functions to resume in a small part of the sea.

In the United States, federal and state agencies gave the go-ahead in 1996 to a large-scale effort to revitalize south Florida's Everglades ecosystem, the famed "river of grass" and treasure trove of wildlife that shrank by half as the natural water system was altered for agricultural and urban expansion (see Chapter 5). At a cost of up to $3 billion over the next 15–20 years, this effort aims to restore the water rhythms of the original natural system. Funding was set back when Florida voters turned down an initiative in November 1996 to impose a penny-a-pound tax on sugar, which would have made the heavily subsidized sugar industry that has caused part of the damage pay for some of the repair.[20]

Like the Aral Sea, the Colorado River delta cannot be returned to anything like it was when Aldo Leopold jour-

---

[19] N.F. Glazovskiy, "Ideas on an Escape from the 'Aral Crisis'," *Soviet Geography*, February 1991.

[20] Nick Madigan, "Environmentalists, Sugar Industry Battle Over Everglades Pollution Figures," *Washington Post*, February 2, 1997.

neyed through it. But at least three areas of critical wetland habitat remain that could be protected or expanded. The largest—the Cienaga de Santa Clara—covers more than 20,000 hectares, harbors a diverse array of species, and may be home to the largest remaining populations of endangered Yuma clapper rail and desert pupfish.[21] Ironically, the Cienaga's water supply is agricultural drainage water sent to the delta by canal from a U.S. irrigation district in Arizona. Protecting it would require the U.S. Department of Interior's commitment to continue supplying this drainage, or other water of sufficient quality, to this part of the delta. Fuller rehabilitation of the delta would require more Colorado River water flowing to it, which in turn would mean changes in how the river is used and managed.[22]

The scientific data do not yet exist, however, to convey to policymakers how much water is needed, and when, in order to increase fish populations, protect endangered species, or achieve other restoration goals. This lack of scientific understanding blocks policy action to protect many candidate ecosystems—and is a critical gap needing to be filled.[23]

Channeling greater investments into the protection and restoration of aquatic ecosystems will take concerted efforts by many people and groups. Scientists and economists have a central role to play in quantifying the economic value of freshwater ecosystem services—and then communicating those values to policymakers. Knowing in advance that the value of a coastal mangrove swamp can reach $300,000 per kilometer for flood con-

---

[21] Edward P. Glenn et al., "Effects of Water Management on the Wetlands of the Colorado River Delta, Mexico," *Conservation Biology*, August 1996.

[22] Jason Morrison, Sandra Postel, and Peter Gleick, *The Sustainable Use of Water in the Lower Colorado River Basin*, (Oakland, Calif.: Pacific Institute for Studies in Development, Environment, and Security, 1996).

[23] Brian Richter, "Ecosystem Level Conservation at the Nature Conservancy: Growing Needs for Applied Research in Conservation Biology," *Journal of the North American Benthological Society*, June 1993.

trol and storm protection alone, for example, might encourage tropical countries to act aggressively to slow the destruction of these coastal jewels.

Unfortunately, many researchers are reluctant to quantify ecosystem values, because the results are rarely defensible as completely accurate. Two adages might apply here, however. First, it is better to be roughly right than precisely wrong. Valuing ecosystem services at zero—which is what happens now by default—is precise, but wrong. Second, bad policy is policy made by others. Unless scientists and economists communicate more often and more effectively with decisionmakers, the choices made are unlikely to be informed or in society's best interest.

Finally, where ecosystems located in poorer countries have global value—because of their species richness, for example—investments from wealthier countries to protect them are not only equitable but sensible. The Global Environment Facility (GEF)—a joint effort of the World Bank, the United Nations Environment Programme, and the United Nations Development Programme—is the primary global financial mechanism set up to transfer funds to developing countries for the incremental costs of protecting global ecological assets.

Although international waters and biodiversity are two of GEF's four areas of focus, only $67 million of its funding allocation to date of $1.4 billion has gone to the protection of global freshwater assets.[24] More than half this freshwater funding is going to Africa's Lake Victoria, where biodiversity and fisheries are greatly threatened. Clearly, broader efforts are needed both through direct bilateral partnerships—as could be developed, for instance, between the United States and Mexico for the Colorado delta—and through a larger

---

[24] Global Environment Facility, *Quarterly Operational Report* (Washington, D.C.: November 1996).

global partnership focused on monitoring, valuing, and safeguarding the planet's vital freshwater ecosystems.

Hydropolitics: From Conflict to Cooperation?

In 1992, when *Last Oasis* was released, I would never have guessed that the potential for "water wars" would begin to diffuse so quickly in so many river basins. In the last few years, countries in several of the major hot spots of water dispute discussed in Chapter 6 have taken strides toward peaceful resolution of their differences. A long road still lies ahead to get to equitable and sustainable water-sharing agreements among all parties in water-scarce river basins. But there is good news to report, and progress to build on.

The most notable achievement comes from South Asia, where India and Bangladesh had long been deadlocked in a dispute over sharing the dry-season flow of the Ganges River. Tensions between the two countries rose to new heights after flows into Bangladesh in 1993 dropped to the lowest level ever recorded, idling irrigation pumps and causing severe crop losses. In October 1995, Prime Minister Begum Khaleda Zia went before the United Nations and called India's heavy river diversions near the border "a gross violation of human rights and justice," and said the Farakka Barrage had become for Bangladeshis "an issue of life and death."[25]

In December 1996, this generation-old dispute dissipated with the signing of a treaty between the two governments on sharing the Ganges at Farakka.[26] The agreement came as a surprise to many water observers, including me, because India—as both the stronger and

---

[25] Cited in Sandra Postel, *Dividing the Waters: Food Security, Ecosystem Health, and the New Politics of Scarcity,* Worldwatch Paper 132 (Washington, D.C.: Worldwatch Institute, September 1996).

[26] "Treaty Between the Government of the Republic of India and the Government of the People's Republic of Bangladesh on Sharing of the Ganga/Ganges Waters at Farakka," December 12, 1996.

upstream country—seemed to have little to gain from bargaining. But new possibilities for progress opened up with political changes in both countries: in India, with the coalition government of H.D. Deve Gowda and its "good-neighbor policy"; in Bangladesh, with the return to power after 21 years of the pro-India party led by Prime Minister Sheikh Hasina.

The new treaty, which is to remain in force for 30 years, grants Bangladesh about three times more Ganges water than it has been receiving in recent years. It establishes a joint commission to monitor flows at Farakka and to carry out the treaty's provisions, and it includes a guaranteed minimum flow to Bangladesh in the event that disagreements arise. Assuming it is ratified, the treaty removes a deep thorn in the diplomatic sides of both countries. While Gowda's government got nothing specific in return for its good will, Bangladesh may see fit to extend the spirit of cooperation with concessions on transit rights, immigration, or other issues important to India.

Comparable leadership has not yet emerged elsewhere, but progress is apparent in other water-scarce river basins. In the Jordan basin, the 1994 peace treaty signed by Israel and Jordan resolved some of their water tensions and included an Israeli commitment to provide an additional 50 million cubic meters of water a year to its neighbor.[27] In the September 1995 interim agreement between Israel and the Palestinians, Israel recognized that Palestinians have rights to West Bank groundwater and agreed that they should get an additional 70–80 million cubic meters of water annually during the interim phase of the peace process. Although not enough to remedy inequities, the added supply should help meet domestic needs.[28]

[27] Jordan-Israel Peace Treaty, *Annex II,Water Related Matters*, October 17, 1994.
[28] Information Division, "Israeli-Palestinian Interim Agreement Annex III— Protocol Concerning Civil Affairs," Israeli Foreign Ministry, Jerusalem, September 1995; Jad Isaac and Jan Selby, "The Palestinian Water Crisis," *Natural Resources Forum*,Vol. 20, No. 1, 1996.

Despite these positive steps, tough challenges remain in the Jordan basin. The issue of equitable allocation of West Bank groundwater—which accounts for about 25 percent of Israel's total nationwide supply—has been left for the "final status" talks between Israel and the Palestinians. Negotiations between Israel and Syria over the Golan Heights, which have broken off temporarily, include a water component that is critical to all parties in the basin. Because Syria is a downstream nation in the Euphrates basin and seeks a larger share of that river's water, it may be inclined to take a more conciliatory approach to sharing the Golan Heights water with downstream Israel, assuming the Golan returns to Syrian control.

Eventually, all five parties in the Jordan basin will likely need to arrive at equitable sharing—and, ideally, joint management—of all the basin's waters. A basinwide water-sharing plan called the Johnston Accord was hammered out in the fifties and accepted by technical experts on all sides, but was never ratified. Given that Israel's current use of Jordan basin water exceeds its Johnston Accord allotment by some 55–75 percent, achieving a fair water-sharing agreement among all parties is unlikely to be easy.[29]

Annual meetings to foster cooperation in the Nile basin have occurred since 1993, and seem to be bearing fruit. At the February 1995 gathering in Tanzania, the water affairs ministers of most of the Nile basin countries—including Egypt and Ethiopia—agreed to form a panel of experts that would be charged with developing a basinwide framework for water sharing aimed at "equitable allocation of the Nile waters."[30] At the February 1997 meeting, the 10

---

[29] Sharif S. Elmusa, *Negotiating Water: Israel and the Palestinians* (Washington, D.C.: Institute for Palestine Studies, 1996).

[30] "Extracts from the Minutes of the 3rd Meeting of the Ministers of Water Affairs in the Nile Basin on Tecconile," and Annex Z, "Project on the Nile Basin Cooperative Framework, Draft Terms of Reference for a Panel of Experts Constituted by the Tecconile Council of Ministers," Arusha, Tanzania, February 9-11, 1995.

Nile basin states (formerly 9) appeared ready to embrace a cooperative approach—in part to secure $100 million in funding for water projects that prospective donors have made conditional on basinwide cooperation.[31] Given Ethiopia's interest in damming upper Blue Nile waters for hydroelectric power and irrigation, these initiatives come none too soon.

Less progress is evident in the other major Middle East hot spot, the Tigris-Euphrates basin. Here, Turkey's position is similar to that of India's on the Ganges: it is both upstream and stronger than the countries that share the river, Syria and Iraq. There seem to be some conciliatory steps toward resolving the dispute over the Euphrates, but a water-sharing agreement is not yet in sight. Meanwhile, Turkey is moving ahead on the large Southeast Anatolia scheme, which will greatly reduce the Euphrates' flow into Syria.

In the Aral Sea basin of central Asia, the presidents of all five former Soviet republics met in January 1994 and approved an action plan for addressing the basin's deteriorating situation. Its centerpiece is a regional water management strategy, drafted in May 1996, that has been agreed to by all five governments. It recognizes the Aral Sea and the two river deltas as "water users" in their own right, deserving of water allocations. It also confirms that principles of international law should apply to decisions of interstate water allocation. Turning these common beliefs into practical policy and a sustainable water-sharing formula, however, will take great effort, given the dire economic and social conditions in the region. A small start was made in February 1997, when the five leaders pledged aid totalling 0.3 percent of their annual fiscal budgets to the International Fund for the Aral Sea, which they

---

[31] Mark Huband, "Nile States Look to New Division of Waters," *Financial Times*, February 27, 1997.

established in 1993.[32]

No enforceable law governs the allocation and use of international waters, but a code of conduct and legal framework based on the principle of "equitable use" is clearly gaining force and legitimacy. As this happens, it will call into question patterns of water allocation that reflect asymmetries of power rather than equity and legitimate need. Events of the last few years may signal that countries now recognize that cooperation will bear more fruit than conflict, and that equitable sharing of water is a prerequisite for a sustainable peace.

Glimmers of a Water Ethic

On the first day of my graduate course on international water issues, I ask students to complete a simple sentence: "Water is ....." When we finish listing as many entries as they can think of, we typically find that responses fall into three broad categories: a fundamental life support, an economic resource, and a source of inspiration and spirituality. As long as water is abundant, these different aspects can comfortably coexist. But as water becomes scarce, they begin to compete. More water devoted to economic activity, for example, may threaten some of its life-support functions or lower its inspirational value. The challenge for society becomes one of managing these competing functions—and staving off conflicts.

Such is the task ahead. The competition among water's diverse roles is here to stay, and the search for the best ways to manage it and minimize conflict is what water policy and planning are now about.

---

[32] Aral Sea Basin Program-Group 1, Interstate Commission for Water Coordination, and World Bank, "Developing a Regional Water Management Strategy: Issues and Work Plan," draft prepared for the Executive Committee of the Interstate Council for the Aral Sea, May 1996; Charles Clover, "Aid Pledged for Shrinking Aral Sea," *Financial Times*, March 3, 1997.

The tools of the trade, many of which are described in Chapter 13, include conservation-oriented pricing, water marketing, regulations, efficiency norms and standards, legal principles, consensus building, and citizen education and participation. Too much is happening in each of these areas to provide even an overview of them in this Introduction to the New Edition. But a few broad trends deserve comment.

First, there is a strong move toward the "commodification" of water—treating it more as an economic good than a gift of nature. In principle, there is nothing wrong with properly valuing water's role as a commodity. Indeed, Chapter 13 underscores how heavy subsidies have discouraged water efficiency, and recommends that water be priced closer to its real cost. Treating water more as an economic good was one of the four principles adopted at a major international water conference in Dublin in 1992.[33] It was echoed in Agenda 21, the plan of action that emerged from the 1992 Earth Summit in Rio de Janeiro, and again in the World Bank's 1993 water policy paper.[34] It is also one of the strategy elements laid out in the just-completed global freshwater assessment requested by the U.N. Commission on Sustainable Development.[35]

The risk, however, is that water's economic functions will be elevated over its life-support functions, and that the three pillars of sustainability—efficiency, equity, and ecosystem protection—will not be given equal weight. One key factor driving the commodification of water is the sheer inability of governments to finance the rising capital, operation, and maintenance costs of irrigation

---

[33] International Conference on Water and the Environment: Development Issues for the 21st Century, *The Dublin Statement and Report of the Conference*, Dublin, Ireland, January 26-31, 1992.

[34] United Nations, *Agenda 21: The United Nations Program of Action From Rio* (New York: U.N. Publications, 1992); World Bank, *Water Resources Management: A World Bank Policy Paper* (Washington, D.C.: 1993).

[35] UN-CSD, *Freshwater Resources of the World*.

and urban water systems. The World Bank has estimated that countries need to invest $600 billion in water infrastructure in just the next decade.[36] In part because of such daunting sums, many governments are turning the construction, operation, management, and sometimes even the ownership of these systems over to the private sector. Although this may help the systems become financially sustainable, there is an inherent risk to the environment and to the poor.

Water "systems" are more than pipes, canals, meters, and treatment works. They can include reservoirs, wetlands, streams, and watershed lands that perform many of the ecosystem services described earlier. Turning control of these natural assets over to a private entity motivated by profit risks the loss of valuable life-support functions. Moreover, there is no guarantee that water systems in private hands will give equity concerns proper weight, since extending coverage to the poor may lower profits.

Although it is too early to judge this trend toward commodification, there is ample reason for heightened vigilance in monitoring it. In the last few years, the privatization of urban water services has greatly picked up speed. Governments have contracted with private companies for the operation of large water systems in Buenos Aires; Dakar, Senegal; Casablanca; Mexico City; Selangor State, Malaysia; and Adelaide, Australia, to name a few. And 30–40 more privatization deals are in the works. Most of the contracts and concessions are going to a handful of French and British companies, leading to a concentration of power and control.

Two industry giants—Lyonnaise des Eaux and Compagnie Générale des Eaux—now supply drinking water to more than 100 million people around the world and account for 20 percent of worldwide corporate R&D

---

[36] Ismail Serageldin, *Toward Sustainable Management of Water Resources* (Washington, D.C.: World Bank, 1995).

spending in the water industry.[37] Given the pace at which privatization is occurring, it seems unlikely that adequate rules and regulations are in place in many cases to protect the resource base, ecosystem services, and the poor.

A parallel trend is under way in irrigated agriculture. Governments in more than 20 countries are decentralizing and privatizing irrigation systems, but in most cases they are turning the systems over to farmers' groups or local organizations, rather than to a private, for-profit company. The Philippines and other Asian countries pioneered farmer participation in small-scale irrigation management during the eighties, but the turnover occurring now often involves large land areas, complex irrigation systems, and a greater degree of privatization.

The Mexican government, for example, turned 2.5 million hectares of irrigated land—78 percent of what is managed federally—over to water user organizations in 1994. Farmer water fees in various districts jumped by 50–180 percent, helping lift the nationwide rate of irrigation financial self-sufficiency from 57 percent to about 80 percent. From a financial standpoint, the strategy thus looks promising, but as a World Bank assessment points out, "it is too early to tell much about the sustainability of the Mexican program and what its long-term efficiency, equity, financial and environmental impacts may be."[38]

A second highly noteworthy trend is the growing interest in returning some water to nature. This reflects rising concern about the loss of water's fundamental life-support functions, and stands as a potentially strong counterweight to the move to treat water as a commodity.

---

[37] "Talking to the Giants," *World Water and Environmental Engineering,* September 1996.

[38] Cecilia M. Gorriz, Ashok Subramanian, and Jose Simas, *Irrigation Management Transfer in Mexico* (Washington, D.C.: World Bank, 1995).

In the United States, river systems and the ecological services they perform are receiving higher priority in decisions concerning irrigation system management, water use permits, hydroelectric dam relicensing, dam operations, water rights, and floodplain management. The U.S. Congress, for example, has voted to decommission two dams in Washington state in order to restore historic salmon runs to the Elwha River.[39] A 1992 federal law dedicates 800,000 acre-feet (987 million cubic meters) of water annually from a large federal irrigation project in California to maintaining fish and wildlife habitat and other ecosystem functions.[40]

In March 1996, the U.S. Bureau of Reclamation released a surge of water from Glen Canyon Dam on the Colorado River in order to mimic the river's pre-dam natural spring flood and thereby restore beaches, sand bars, and critical habitat in the Grand Canyon.[41] And on the heels of a final decision in 1994 to protect California's Mono Lake by reducing Los Angeles's diversions from streams in the basin (see Chapter 13), two more court decisions relying on the public trust doctrine—in Idaho and Washington—have resulted in water being reserved for natural ecosystems.

The unsung heroes behind many of these victories are NGOs and citizens' groups, which have emerged as a powerful new force for conservation and ecosystem protection. In Massachusetts, for instance, the Hadley-based Water Supply Citizens Advisory Committee helped defeat a dam proposed for an expansion of Boston's water supply; it also worked to launch a conservation effort that has made the expansion unneces-

---

[39] Shawn Cantrell, "Restoring the Bounty (and the Beauty) of the Elwha," *Living Oceans News*, Fall 1996.

[40] Brian Gray, "The Modern Era in California Water Law," *Hastings Law Journal*, January 1994.

[41] Jeff Hecht, "Grand Canyon Flood a Roaring Success," *New Scientist*, August 31, 1996.

sary by reducing water use in greater Boston about 25 percent.[42] In California, much of the credit for the Mono Lake victory goes to the Mono Lake Committee, which tirelessly pursued the protection of this unique ecosystem through litigation, legislation, and public education. Elsa Lopez, head of Mothers of East Los Angeles, shows in the "Last Oasis" documentary how a water conservation program can not only save water but also create jobs and pride in the local community. Such groups and individuals are living proof of the water ethic at work.

A third trend has to do with the widening gap in water use between rich and poor—both within and between countries. The finite nature of water and its approaching limits inexorably link the water demands of the economically privileged and the disadvantaged. The billions of poor people eating low on the food chain, for example, allow the high-meat diets of the wealthy to persist, for there is not enough land and water for all to have meat-intensive diets. There is only so much ecological space, and as it fills up, some people can consume more than their fair share only by others consuming less than theirs. The persistence of such inequity is a sure recipe for social tension and conflict.

In many developing countries, for instance, demand for amenities such as golf courses is rising as some people become richer. Malaysia, Thailand, Indonesia, South Korea, and the Philippines have 550 golf courses in all, with plans for 530 more on the drawing boards.[43] Even Egypt—which is growing by 1 million people every nine months and is nearly at the limit of its available water—has plans for eight new golf courses.[44] These countries

---

[42] Amy Vickers, Amy Vickers & Associates, private communication, Amherst, Mass., March 1997.

[43] Philip Shenon, "Fore! Golf in Asia Hits Environmental Rough," *New York Times*, October 22, 1994.

[44] John Lancaster, "Desert Turns to Greens as Egypt's Affluent Revive Colonial Links," *Washington Post*, February 22, 1997.

seem willing to gamble that economic growth will pay for future grain imports, that such imports will be available, and that providing water to golf courses for a few while many others lack safe drinking water will not destabilize social relations.

No country has put into effect a water strategy that is firmly founded on the pillars of sustainability, although South Africa is attempting to do so. In early 1996, Kader Asmal, the Minister of Water Affairs and Forestry, laid out principles for a fundamental overhaul of the nation's water law and management. The top priorities are providing each South African with access to at least 25 liters of water a day for meeting basic needs, allocating water to the environment to prevent the loss of ecosystem functions, pricing water at levels that reflect its value, mandating that water suppliers adopt conservation measures, and reserving water for countries downstream in order to promote regional cooperation.[45]

Greater equity in access to water is the centerpiece of the strategy. In fact, the revised constitution includes the right of all citizens to have access to sufficient water. As Asmal said in a 1996 speech, "In South Africa, it is precisely because we have put peoples' immediate needs first that we have gained the ethical basis to demand a long term view of sustainability."[46]

One final thought. As I write this Introduction, a bill has been introduced into the U.S. Congress to repeal the national water-efficiency standards for household plumbing fixtures that were passed in late 1992. This brings home the point that no victory is ever secure—no

---

[45] "Water Law Review Supported By All," draft press release, Ministry of Water Affairs and Forestry, Pretoria, South Africa, February 6, 1996; "A Step Towards Equity in Access and Optimal Use of Water," *Water Sewage and Effluent*, March 1996; Eddie Koch, "A Watershed for Apartheid," *New Scientist*, April 13, 1996.

[46] Kader Asmal, Minister of Water Affairs and Forestry of South Africa, Speech delivered at the First Meeting of the Global Water Partnership, Stockholm, Sweden, August 9, 1996.

matter how hard it was to win. On the path to a sustainable water future, constant vigilance is required to consolidate gains and keep moving forward.

Glimmers of the water ethic are now visible. To make it take hold and flourish, we may have to tap a trait that water itself embodies—synergy. Who would have believed that combining two parts of hydrogen gas with one of oxygen would produce this amazing substance called water? That it does is a useful reminder that a synergistic mix of people's efforts may produce miraculous results.

Sandra Postel

Global Water Policy Project
Amherst, Massachusetts
Connecticut River Watershed

March 1997

Acknowledgments

The seeds of this book were sown more than a decade ago, when as a consultant I had the opportunity to consider how the desert city of Tucson, Arizona, could reduce its claim on the water bodies it was depleting and overusing. During my nine years at Worldwatch Institute, I extended that interest to the world at large, and became more deeply aware of water's sustaining role in the natural environment.

I have written this book to share not only what I have learned, but also the experiences and successes of many people who have grappled with water scarcity problems and solutions. Writing it has also been an outlet for my personal concerns about the consequences of continuing down the path we are on. My aim is to draw attention to a sleeper of a problem—the escalating pressures we are placing on water systems, on our rivers, lakes, wetlands, and underground aquifers. We have so taken water for granted that our use and abuse of it now threaten to undermine the very life-support systems we depend on.

I sincerely thank the Ford Foundation for sharing these concerns with me enough to support the research, writing, and marketing of this book with a generous grant. Coming from a foundation with a long and impressive track record dealing with global water issues, the vote of confidence this represented meant a great deal. I hope our common aims were well served.

I am indebted to many people who played a part in the creation of this book. As my research assistant throughout the project, Peter Weber contributed his insights and research skills, as well as suggestions that markedly improved the book. I greatly appreciated his selfless involvement in all the project's tasks—from data-gathering and analysis to reviewing numerous drafts and checking facts. Heather Hanford supported the information-assembling, documentation, and organizational aspects of this project with her usual degree of efficiency and reliability, even while fulfilling her many other duties as Worldwatch librarian. *Last Oasis* is measurably better for the efforts both Peter and Heather put into it.

More than a dozen people inside and outside of the Institute took time out of busy schedules to review all or part of an early draft of the manuscript. I thank my Worldwatch colleagues Lester Brown, Alan Thein Durning, Christopher Flavin, Nick Lenssen, and John Ryan for doing so. Their comments helped me iron out the wrinkles and hone the book's message.

Last Oasis also benefited from the comments and critiques of eight specialists who have more experience than I do in specific areas of water management. Of course, any errors of fact or interpretation that remain are my responsibility. For sharing their time and expertise with me, I thank John Ambler, program officer with the Ford Foundation in New Delhi, India; Saul Arlosoroff (along with his assistant, Jodi Felberg), project manager in water and sanitation at the World Bank; E. Walter Coward, Jr., director of the Rural Poverty and Resources Program at the Ford Foundation in New York; Peter Gleick, director of the Global Environment Program at the Pacific Institute for Studies in Development, Environment, and Security in Oakland, California; Ruth Meinzen-Dick, research fellow at the International Food Policy Research Institute in Washington, D.C.; Deborah Moore, staff scientist at the Environmental Defense Fund in Oakland, California; Frederik van Bolhuis, environmental economist with the Global Environment Facility at the World Bank; and Amy Vickers, principal of Amy Vickers & Associates in Boston, Massachusetts. To Frederik and Amy, I offer special thanks for unwavering friendship and support during the trying times that inevitably arise in writing a book.

Numerous people helped make my research travels more productive and enjoyable. Here, for their generous hospitality, I thank Luis Manuel Guerra in Mexico City, Nicholas Hopkins in Cairo, and, in Israel, Danny Sherban, Uri Or, and Rachel Guy.

Finally, I must also extend thanks to a few people from my past—professor Thomas Gerrard, whose love of geology and skill at teaching it piqued my interest in the workings of the natural world, engineer Bill Betchart, who guided me through my early water conservation analyses, and planner Carl Grove who, unbeknownst to him, taught me a great deal about solving problems.

Finally, my deepest appreciation goes to the person who has known me longer than any other and supported me at every step. It might just be that our many summer days by the ocean, an early-adolescent pilgrimage to Walden Pond, and our restorative vacation beside Lake Champlain had something to do with my choosing a path that led to the writing of *Last Oasis*. Certainly, his confidence in me did. With warm memories and thanks, I dedicate this book to Harold Postel, my father.

Sandra Postel

Worldwatch Institute
1776 Massachusetts Ave., N.W.
Washington, D.C. 20036

July 1992

Last Oasis

1

An Illusion
of Plenty

Life could hardly be more different in the east African town of Lodwar, Kenya, and the western U.S. metropolis of Phoenix, Arizona. At the touch of a tap, a child in Phoenix has ample water for drinking, bathing, even swimming in a backyard pool. His family probably uses some 3,000 liters of water on a typical day, enough to fill their bathtub 20 times over. A child living on the outskirts of Lodwar, on the other hand, daily treks several hours to a well or spring to help her mother bring home a couple of jugs of water. Her family uses barely 5 percent as much water as the Phoenix household, just enough to satisfy their most basic needs.

Yet when it comes to the amount of water nature makes available, Lodwar and Phoenix are "sister cities." Each gets a meager 16–18 centimeters of rainfall a year.

And in each place human numbers have outstripped the ability of local water supplies to sustain a moderate standard of living.[1]

Lodwar and Phoenix exhibit two very different faces of water scarcity. In Lodwar, people experience scarcity in its rawest form—it adds drudgery and insecurity to their everyday lives. And their plight is exacerbated, as it is in many developing regions, by the poor's lack of access to even the limited supplies available. In Phoenix, however, scarcity is masked by the damming, diverting, and pumping of water from near and far to make the city not only livable but lush. An illusion of plenty has been created in water-scarce Phoenix—which leads to overconsumption and adverse consequences for the environment and for future generations.

In a sense, masking scarcity is a principal aim of water development, the collection of engineering projects and technologies that give people access to and control over nature's supply. But all too often it has proceeded without regard for harmful side-effects. We build ever more and larger projects to meet spiraling demands wherever they arise, but pay little mind to the ecological services of rivers, lakes, and wetlands that are lost in the process. And we deplete groundwater reserves to meet today's needs and desires, with no thought of the consequences for generations down the line.

For most of us, water scarcity conjures up visions of drought, those temporary dry spells that nature inflicts from time to time. But while droughts capture headlines and grab our attention, the far greater threat posed by our escalating water consumption goes largely unnoticed. In many parts of the world, water use is nearing the limits of natural systems; in some areas, those limits have already been surpassed. A number of areas could enter a period of chronic shortages during this decade, including much of Africa, northern China, pockets of India, Mexico, the Middle East, and parts of western North America.

Signs of water stress abound. Water tables are falling, lakes are shrinking, and wetlands are disappearing. Engineers propose "solving" water problems by building mammoth river diversion schemes, with exhorbitant price tags and untold environmental effects. Around Beijing, New Delhi, Phoenix, and other water-short cities, competition is brewing between city-dwellers and farmers who lay claim to the same limited supply. And people in the Middle East have heard more than one leader voice the possibility of going to war over scarce water.

In the quest for better living standards and economic gain, modern society has come to view water only as a resource that is there for the taking, rather than a living system that drives the workings of a natural world we depend on. Harmonizing human needs with those of a healthy environment will require new ways of using and managing water. And it will require adjusting our production and consumption patterns so as to remain within ecological limits.

In each major area of water use—agriculture, industry, and cities—demands have risen markedly since 1950. At that time, both population and material consumption began a steep climb, driving water use rapidly upward. By and large, those pressures continue today, as worldwide needs for food, industrial products, and household services expand.

Agriculture claims the lion's share of all the water taken from rivers, lakes, and aquifers, accounting for an estimated 65 percent of global water use. As opportuni-

ties to extend cropland area have dwindled, augmenting
food production has come to depend more on coaxing
higher yields from existing farmland, which often requires irrigation. Over the course of this century, as the
number of people to feed swelled from 1.6 billion to
more than 5.4 billion, agriculture's water use increased
fivefold. The really rapid rise began around mid-cen-
tury, when water development entered its heyday, and
continued as the Green Revolution—involving fertilizers and pesticides, high-yielding seeds, and irrigation—
took hold and spread.[2]

Industries make the second largest claim on the
world's water bodies, accounting for a fourth of global
water use. Generating electricity in thermal power
plants (nuclear and fossil fuel) takes copious amounts of
water, as does making the paper, steel, plastics, and
other materials we use every day. Spurred by droughts
and strict pollution control requirements, industries in
the richer countries have shown that they can reduce
their water use dramatically by recycling and reusing
their supplies. Yet these technologies remain greatly un-
derused, particularly in the developing world, where industry's water use is now rising rapidly.[3]

Water deliveries to households, schools, businesses,
and other municipal activities account for less than a
tenth of global water use today. Nonetheless, meeting
these needs is no easy task. Drinking water must be
treated to a high level of quality and supplied with a high
degree of reliability, which makes it expensive. As cities
expand, planners reach out to capture ever more distant
and costly sources. Tapwater in many homes in Los Angeles, for instance, originates hundreds of kilometers
away in northern California or the Colorado River
basin. By the end of this decade, some 22 cities world-

wide will have populations of 10 million or more, and 18 of them will be in the Third World. Serving these dense population centers will in many cases take more water, capital, and energy than is available or affordable.[4]

Already today, there remains a large unmet demand for household water. Nearly one out of every three people in the developing world—some 1.2 billion people in all—do not have access to a safe and reliable supply for their daily needs. Often they resort to shallow wells or stagnant pools that are easily contaminated with human and animal waste. As a result, waterborne diseases account for an estimated 80 percent of all illnesses in developing countries. And women and children walk several kilometers each day just to collect enough water for drinking, cooking, and cleaning, a drudgery that saps time and energy from more productive activities.[5]

Added up, total human demands for water—including agriculture, industries, and cities—still seem comfortably below the amount nature makes available each year. But this, too, is illusory. Much rainwater runs off in floods, falls in places too remote for us to capture it, or is needed to support the myriad other species and ecosystems with which we share the planet, and on which we depend.

Moreover, in many places, pollution is rapidly diminishing the usable supply. Each liter of polluted wastewater contaminates many additional liters in the water body that receives it. In Poland, for example, the share of river water of highest quality for drinking has dropped from 32 percent to less than 5 percent during the last two decades. Some three quarters of that nation's river water is now too contaminated even for industrial use. Similar situations increasingly can be found in developing countries, where unchecked pollution poses a mounting threat during industrialization.[6]

Although water is part of a global system, how it is used and managed locally and regionally is what really counts. Unlike oil, wheat, and most other important commodities, water is needed in quantities too large to make it practical to transport long distances. No global water crisis is likely to shake the world the way the energy crisis of the seventies did. But with key crop-producing regions and numerous metropolitan areas showing signs of water scarcity and depletion, global food supplies and economic health are in jeopardy. Moreover, global warming from the buildup of greenhouse gases could greatly complicate regional water problems by shifting the patterns of rainfall and runoff that agriculture and urban water systems are geared to.

Without question, water development has been a key to raising living standards, and it needs to be extended to the one fifth of humanity who have largely missed out on its benefits. But, as Part I of this book will show, in our rush for economic growth, food sufficiency, and material well-being we have repeatedly ignored nature's limits—depleting underground aquifers, deforesting watersheds, and diminishing streamflows to ecologically damaging levels.

Achieving water balance will not be easy. The policies, laws, and practices that shape water use today rarely promote all three basic tenets of sustainable resource use—efficiency, equity, and ecological integrity. Even a casual glimpse around the world shows water allocation and use to be in a chaotic state. While farmers in California's Central Valley were spreading copious amounts of inexpensive irrigation water on cotton and rice, Los Angeles was draining the streams feeding fragile Mono Lake to fill swimming pools and wash cars.

Sugarcane growers in the Indian state of Maharashtra take 50 percent of available irrigation supplies even though they occupy only 10 percent of the cropland. And from the Everglades to the Aral Sea, aquatic habitats unravel from the siphoning off and pollution of rivers and streams.[7]

Taking heed of water's limits, and learning to live within them, amounts to a major transformation in our relationship to fresh water. Historically, we have approached nature's water systems with a frontier philosophy, manipulating the water cycle to whatever degree engineering know-how would permit. Now, instead of continuously reaching out for more, we must begin to look within—within our regions, our communities, our homes, and ourselves—for ways to meet our needs while respecting water's life-sustaining functions.

Doing more with less is the first and easiest step along the path toward water security. By using water more efficiently, we in effect create a new source of supply. Each liter conserved can help meet new water demands without damming another stretch of river or depleting more groundwater. With technologies and methods available today, farmers could cut their water needs by 10–50 percent, industries by 40–90 percent, and cities by a third with no sacrifice of economic output or quality of life. As Part II will show, most investments in water efficiency, recycling, reuse, and conservation yield more usable water per dollar than investments in conventional water supply projects do. But they will not materialize until policies, laws, and institutions begin to foster such measures rather than hinder them, as discussed in Part III.

New technologies and better policies have much to offer toward the goal of achieving a secure water future, but they will take us only so far. They alone cannot avert conflicts and shortages where populations are expanding faster than efficiency measures can release new supplies. Any hope for balancing the water budgets of most Middle Eastern countries, for instance, rests as much on lowering birth rates as it does on modernizing irrigation systems. And in many water-short African countries, slowing population growth appears to be the only way of meeting minimal per capita needs in the near future.

A new water era has begun. In contrast to earlier decades of unfettered damming, drilling, and diverting to gain ever greater control over water, the next generation will be marked by limits and constraints—political, economic, and ecological. Yet numerous opportunities arise as well. Exploiting the market potential of new water-saving technologies is an obvious one. And in many cases, achieving better water management will require decentralizing control over water, and moving from top-down decisionmaking to greater people's participation—a shift necessary for better human and economic development overall.

Most fundamentally, water scarcity challenges us to adopt a new ethic to guide our relationship to the earth's natural systems, to other species, and to each other. Recognizing ourselves as part of the life-support network we depend on and learning to live within water's limits are integral aspects of creating a society that is sustainable in all respects. Measures to conserve water and use it more efficiently are now the most economical and environmentally sound water supply options available for much of the world—and they have barely been tapped. Together, they constitute our "last oasis."

I

Trouble on Tap

The World's Renewable Water Supply

Annual Precipitation Minus Evaporation

Driest   ☐   0-50 millimeters

         ☐   50-500 millimeters

         ▨   500-1,000 millimeters

Wettest   ■   More than 1,000 millimeters

Source: Based on State Hydrological Institute of the U.S.S.R.,
Atlas of World Water Balance
(Paris: UNESCO, 1977).

2

Signs
of Scarcity

Having viewed the earth from space, we know it is a strikingly blue planet. It is hard to believe scarcities could arise in the midst of such amazing water wealth. The total volume of water, some 1,360,000,000 cubic kilometers, would cover the globe to a height of 2.7 kilometers if spread evenly over its surface. But more than 97 percent is seawater, 2 percent is locked in icecaps and glaciers, and a large proportion of the remaining 1 percent lies too far underground to exploit.[1]

Fortunately, a tiny fraction of the planet's water is renewed and made fresh by nature's solar-powered water cycle. Each year, evaporation fueled by the sun's energy lifts some 500,000 cubic kilometers of moisture into the atmosphere—86 percent from the oceans and 14 percent from the land. An equal amount falls back to earth as rain, sleet, or snow, but it is distributed in different proportions: whereas the continents lose about 70,000 cubic kilometers through evaporation, they gain 110,000 through precipitation.[2]

As a result, roughly 40,000 cubic kilometers are transferred from the sea to the land each year. This constitutes the world's renewable freshwater supply—that which can be counted on year after year. At today's population size, it amounts to an annual average of about 7,400 cubic meters per person, several times what a society needs to support a moderate standard of living for its people. But this water is distributed very unevenly (see map, p.26), and not all of it can be used by humans as it makes its way back to the sea. Two thirds runs off in floods, leaving about 14,000 cubic kilometers as a relatively stable source of supply. And protecting wetlands, deltas, lakes, and rivers, as well as safeguarding water quality, requires that a substantial share of this stable flow be left alone to run its natural course.[3]

Although water is a renewable resource, it is also a finite one. The water cycle makes available only so much each year in a given location. That means supplies per person, a broad indicator of water security, drop as population grows. Thus per capita water supplies worldwide are a third lower now than in 1970 due to the 1.8 billion people added to the planet since then.[4]

One of the clearest signs of water scarcity is the increasing number of countries in which population has surpassed the level that can be sustained comfortably by the water available. As a rule of thumb, hydrologists designate water-stressed countries as those with annual supplies of 1,000–2,000 cubic meters per person. When the figure drops below 1,000 cubic meters (2,740 liters per person a day), nations are considered water-scarce—that is, lack of water becomes a severe constraint on food production, economic development, and protection of natural systems.[5]

Today, 26 countries, collectively home to 232 million people, fall into the water-scarce category. Many of them have very high population growth rates, and so their water problems are deepening fast. (See Table 2–1.) For a country like Egypt, which gets practically no rainfall, water flowing in from upstream neighbors is a precious lifeline. Africa has the largest number of water-scarce countries, 11 in all, and by the end of this decade, four others will join the list. By 2000, the total number of Africans living in water-scarce countries will climb to 300 million, a third of the continent's projected population.[6]

Nine out of 14 countries in the Middle East already face water-scarce conditions, making it the most concentrated region of water scarcity in the world. Populations in six of them are projected to double within 25 years, and so a rapid tightening of supplies is inevitable. With virtually all Middle East rivers being shared by several nations, tensions over water rights are a potent political force throughout the region, and could ignite during this decade. (See Chapter 6.)[7]

Although the population-water equation suggests where to expect trouble, numerous physical symptoms of water stress already exist—and not just in water-scarce countries, but in parts of water-wealthy ones as well. Among the most pervasive problems is that of declining water tables, which is caused by using groundwater faster than nature replenishes it. If pumping is not brought into balance with recharge, eventually the underground supply becomes too expensive to keep tapping, too salty to use as it is pulled up from greater

TABLE 2-1. *Water-Scarce Countries, 1992*[1]

Region/Country	Renewable Water Supplies	Population	Population Doubling Time
	(cubic meters per person)	(million)	(years)
Africa			
Algeria	730	26.0	27
Botswana	710	1.4	23
Burundi	620	5.8	21
Cape Verde	500	0.4	21
Djibouti	750	0.4	24
Egypt	30	55.7	28
Kenya	560	26.2	19
Libya	160	4.5	23
Mauritania	190	2.1	25
Rwanda	820	7.7	20
Tunisia	450	8.4	33
Middle East			
Bahrain	0	0.5	29
Israel	330	5.2	45
Jordan	190	3.6	20
Kuwait	0	1.4	23
Qatar	40	0.5	28
Saudi Arabia	140	16.1	20
Syria	550	13.7	18
United Arab Emirates	120	2.5	25
Yemen	240	10.4	20
Other			
Barbados	170	0.3	102
Belgium	840	10.0	347
Hungary	580	10.3	—
Malta	80	0.4	92
Netherlands	660	15.2	147
Singapore	210	2.8	51
Total Population		231.5	

[1]Countries with per capita renewable water supplies of less than 1,000 cubic meters per year. Does not include water flowing in from neighboring countries.

SOURCE: See endnote 6.

depths, or simply too depleted to serve as a supply. Overuse of groundwater is now ubiquitous in parts of China, India, Mexico, Thailand, the western United States, north Africa, and the Middle East.

Some of the most worrisome cases of unsustainable groundwater use involve "fossil" aquifers, underground reservoirs that hold water hundreds or thousands of years old and that receive little replenishment from rainfall today. Like oil reserves, these aquifers are essentially nonrenewable: pumping water from them depletes the supply in the same way that extractions from an oil well do. Farms and cities that depend on this water will eventually face the problem of what to do when the wells run dry.

The arid kingdom of Saudi Arabia represents one of the most egregious cases of unsustainable water use in the world today. This Persian Gulf nation now mines fossil groundwater for 75 percent of its water needs, and that dependence is growing. Groundwater depletion has been averaging about 5.2 billion cubic meters a year, and the rate is projected to increase by nearly half during the nineties.[8]

A major push by the Saudi government to raise food self-sufficiency largely explains the escalation in groundwater pumping. By heavily subsidizing land, equipment, and irrigation water, and by buying crops at several times the world market price, the government has encouraged large-scale wheat production in the desert. Though the kingdom imports barley and other food crops, it became self-sufficient in wheat in 1984, and has since joined the ranks of the world's top wheat exporters. In early 1992, King Fahd authorized payments totaling $2.1 billion for 1991's record 4-million-ton wheat crop, which he could have purchased on the world market for one fourth the price.[9]

Since Saudi crop production depends on fossil groundwater, little of this grain can be considered a reliable portion of the long-term food supply—either for the Saudis or for those countries receiving its exports. At the depletion rate projected for the nineties, and assuming 80 percent of the groundwater reserve can be exploited, the supply would be exhausted in 52 years. At the faster extraction rates projected for 2000–10, the reserves would dry up much sooner. And even before that happens, the groundwater will likely become too salty to use without expensive treatment.[10]

A similar scenario is unfolding in the north African nation of Libya, another water-scarce country. In late August 1991, with great pomp and ceremony, Colonel Muammar Qaddafi christened "The Great Man-Made River Project," one of his dream-come-true ventures and the world's largest civil engineering effort during the last seven years. It consists of a giant pipeline that will ship water from Libya's southern desert north toward the Mediterranean coast, where overpumping has already diminished local resources and caused saltwater to invade the freshwater supply.[11]

As in Saudi Arabia, the Libyans are linking their economic fate to a nonrenewable water supply. The desert aquifer filled with water some 30,000 years ago, when north Africa received much more rainfall than it does today. With completion of the project's first phase, 730 million cubic meters per year will be mined from the underground reserve and sent north. Ultimately, if all five of the planned phases are completed (at an estimated total cost of $25 billion), the pipeline network will transport as much water as a fairly good-sized river, and the pace of groundwater mining will greatly quicken.[12]

Engineers project that the wells will run dry within 40–60 years, leaving the farms, industries, and people supported by this water with a highly uncertain future. Generations down the line will no doubt question the wisdom of spending one-time oil earnings on the irreversible depletion of a finite water reserve in order to grow crops Libya could have bought far more cheaply in a global marketplace that has grain to spare. In a few decades, when food and water to grow it are both more scarce, the water reserve will be dry and Qaddafi's venture will seem an ill-conceived extravagance.[13]

In the United States, a large and important aquifer system in the High Plains, which contains the well-known Ogallala formation, has been undergoing depletion for several decades. Stretching from southern South Dakota to northwest Texas, the High Plains aquifer supplies about 30 percent of the groundwater used for irrigation in the United States.[14]

The most severe depletion has occurred in northwest Texas, where heavy pumping for irrigation began to expand rapidly in the forties. As of 1990, 24 percent of the Texas portion of the Ogallala had been depleted, a loss of 164 billion cubic meters—equal to nearly six years of the entire state's water use for all purposes. As pumping costs rose and irrigation became uneconomical, the irrigated area in northwest Texas shrank rapidly, falling from a peak of 2.4 million hectares in 1974 to 1.6 million hectares in 1989, a drop of one third. With the High Plains accounting for 65 percent of Texas's total irrigated area, statewide irrigation trends mirror this decline. (See Figure 2–1.)[15]

The economy of northwest Texas is centered largely

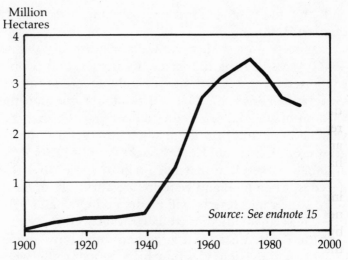

FIGURE 2-1. *Irrigated Area in Texas, 1900–90*

on grain-fed beef production. With a fourth of the region's water supply already depleted, the long-term viability of that industry is in question. Much of the land is now in dryland production and yields far less than when it was irrigated. Recognizing the dire consequences of unchecked groundwater depletion, High Plains water officials and farmers have turned to conservation and efficiency on a large scale, and the rate of aquifer depletion has slowed. (See Chapter 8.) Nonetheless, it remains another example of the shortsighted draining of a critical water reserve—in this case, to irrigate crops destined mainly for cattle.

In many regions, as demands continue to rise and as water supply projects get more difficult to build, water budgets are becoming badly imbalanced. China—with 22 percent of the world's people and only 8 percent of its fresh water—faces obvious water constraints. The nation's predicament is particularly severe in and around Beijing, the important industrial city of Tianjin, and other portions of the North China Plain, a vast expanse of flat, fertile farmland that yields a quarter of the country's grain. Water tables beneath the capital have been dropping 1–2 meters a year, and a third of its wells have reportedly gone dry. All told, some 100 Chinese cities and towns, mostly in the northern and coastal regions, have suffered shortages in recent years.[16]

Even as supplies get tighter, water demands are growing rapidly in north China. Planners project that water needs in the central urban district of Beijing will increase by 38 percent over the next decade, while demands in the surrounding rural district are expected to rise 12 percent. That sets the stage for Beijing's total water demand in 2000 to outstrip currently available supplies by 70 percent. For Tianjin, even optimistic projections of the water balance by decade's end show that city heading for a shortfall of 36 percent.[17]

In some cases, water problems stem directly from mismanagement and degradation of the land. When rain hits the earth, it either runs off immediately into rivers and streams to head back to the sea, soaks into the land to replenish soil moisture and groundwater supplies, or is evaporated or transpired (by plants) back into the atmosphere.

Land degradation, whether from deforestation, overgrazing, or urban development, shifts the proportion of rainfall following each of these paths. With reduced vegetative cover and soils less able to absorb and hold water, degraded land increases flash runoff and decreases seepage into the soil and aquifer recharge. As a result, less soil moisture and groundwater are available to draw upon during the dry season, and during the rainy season the rapid runoff intensifies flooding and soil erosion.

For a country like India, which gets 80 percent of its rainfall in three to four months, with much of it coming in just a few monsoon storms, the ability of the land to absorb water and moderate its release can mean the difference between devastating floods and droughts and a manageable year-round supply. According to Jayanto Bandyopadhyay of the Research Foundation for Science and Ecology in Dehra Dun, the loss of aquifer recharge from land degradation largely explains why high-rainfall areas in India now petition for drought relief. Tens of thousands of villages across the subcontinent experience shortages, and their numbers are growing. Says Bandyopadhyay, "Water is only a renewable resource if we respect the ecological processes that maintain and give stability to the water cycle. This, India has signally failed to do."[18]

The disruption of critical watershed functions has combined with rapidly rising demands fueled by population growth, irrigation expansion, and industrial development to compound pressures on India's water bodies. Groundwater overdrafting is now widespread in parts of India's Deccan Plateau, including the states of Andhra Pradesh, Karnataka, Maharashtra, and Tamil Nadu. During recent decades, increased pumping has been accompanied by deterioration of the centuries-old system of irrigation by tanks—small reservoirs that both store water and help replenish underlying aquifers. The result has been extensive groundwater depletion.[19]

Between 1946 and 1986, for example, the water table in parts of Karnataka dropped from 8 meters below the surface down to 48. Just in the two decades prior to

1986, a period when groundwater wells proliferated, the area of overdrafting in Karnataka expanded ninefold. Foreshadowing increased competition and conflict over water, researchers M.G. Chandrakanth and Jeff Romm conclude that "allocating groundwater between uses, between users, between areas of origin and destination, and between generations is becoming a critical problem in the Deccan states."[20]

These examples by no means constitute a complete cataloging of the water problems evident in the world today; many others will come to light in later chapters. But together they illustrate some of the clearest signals of water stress. Shrinking groundwater reserves, falling water tables, increased flooding and droughts, and water budgets that are badly out of balance are tangible indications that the imperatives of efficiency and ecological integrity have been ignored. Failing to heed these warnings of unsustainable water use will make the consequences hit home sooner and more harshly.

3

Engineering's Promise

Parakrama Bahu the Great, twelfth-century king of what is now Sri Lanka, set forth the ultimate challenge for water engineers: "Let not even a small quantity of water obtained by rain go to the sea, without benefiting man."[1]

Eight hundred years later, Parakrama's dictum has been carried out to a remarkable extent in many parts of the world. Engineers have built more than 36,000 large dams to control floods and to provide hydroelectric power, irrigation, industrial supplies, and drinking water to an expanding global population and economy. Today, construction begins on an average of 170 dams around the world each year. Rare is the river that now runs freely toward the sea, and many that still do are slated to come under control soon. Of Japan's 109 major rivers, for instance, only one—the Nagara—remains undammed, and construction works to tame it are well under way.[2]

This engineering frenzy has embodied the hope that, by controlling an ever greater portion of nature's water cycle, humanity could be freed from the constraints posed by rainfall's unequal distribution in place and time. Food production, industrial output, and urbanization all expanded with the aid of large reservoirs to store runoff and long canals to transfer water from one place to another. Technological feats such as Egypt's Aswan High Dam and the California Aqueduct have literally made deserts bloom. With the skills of modern engineering, almost no place seemed too parched or remote to reach with a lifeline.

As a result, water supply and demand have risen steadily. Global water use has tripled since 1950, and now stands at an estimated 4,340 cubic kilometers per year—eight times the annual flow of the Mississippi River. (See Figure 3–1.) This total, which includes only what is removed from surface and underground waters, amounts to some 30 percent of the world's stable renewable supply. But we actually make use of a far larger share, since water bodies dilute pollution, generate electricity, and harbor fish and wildlife. And because of improved living standards, world water demand has been growing faster than population: at 800 cubic meters, per capita use today is nearly 50 percent higher than it was in 1950, and in most of the world continues to rise.[3]

But limits to Parakrama's vision of an ever-expanding supply are swiftly coming to light. Engineers naturally first selected the easiest and least-costly sites for water development. Over time, water projects have become increasingly complex, expensive to build, and more

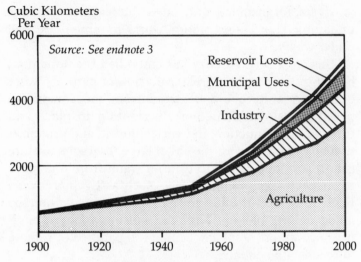

FIGURE 3-1. *Estimated Annual World Water Use, Total and by Sector, 1900–2000*

damaging to the environment. Fewer dams and diversion projects are making it off the drawing boards, and most of those that do will deliver water at a far higher price than in the past.

Worldwide, the rate of dam construction during the last decade has averaged less than half that of the preceding 25 years—170 annually, compared with some 360 per year from 1951 to 1977. In Australia, North America, and Western Europe, few affordable and acceptable sites remain for damming and diverting more river water. Reservoir storage capacity in the United States has plateaued, having increased little over the last decade. The Central Utah Project, serving the Salt Lake City area, and the Central Arizona Project, supplying Phoenix, Tucson, and neighboring towns and cities with Colorado River water, represent the last of the large federal water projects. As a U.S. Congressional Budget Office report flatly concluded nearly a decade ago: "The days of huge federal outlays for equally large water projects appear to be over."[4]

An even louder death knell sounded with the 1990 decision of the U.S. Environmental Protection Agency (EPA) to veto the Two Forks Dam, a $1-billion project that would have flooded Colorado's scenic Cheesman Canyon on the South Platte River in order to increase water supplies for Denver and surrounding communities. Although no federal funds were involved, EPA Administrator William K. Reilly used his review authority under the U.S. Clean Water Act to make it clear that environmental values would no longer be wantonly sacrificed at the altar of water development. Several special districts and a private water company have filed a lawsuit contesting EPA's action, though Denver itself has not. The Two Forks veto casts doubt on the future of other projects that would destroy unique natural areas prized for their wilderness, habitat, recreational, or aesthetic values.[5]

As cities and cropland have expanded in dry regions of the world, planners and engineers have looked farther afield for new water sources to tap. A number of massive water transfer schemes have been proposed in response to threats of regional water scarcity. Most were conjured up during the fifties and sixties, a time of relatively cheap energy, inexpensive capital, and minimal concern about environmental damage. With one possible exception in China (discussed later in this chapter), none of these projects is moving forward. Yet they periodically resurface, perpetuating the notion that more gargantuan engineering feats can solve our water problems.

The former Soviet Union has come close to pursuing

two of the largest river diversions ever envisaged. One would transfer water from northern European lakes and rivers into the Volga River basin, primarily to help stabilize the level of the Caspian Sea, which had declined steadily from the thirties until the late seventies. The other scheme, now almost legendary, would reverse a portion of northward-flowing Siberian rivers in order to channel water to the parched lands of central Asia, home of the shrinking Aral Sea. As recently as the mid-eighties, both projects looked likely to move forward. Indeed, in late 1984, the government gave the go-ahead for the first stage of the European rivers diversion, and construction of access roads, worker housing, and other infrastructure began.[6]

With the ascendancy of Mikhail Gorbachev to the presidency in 1985 and the new policy of glasnost, a group of influential writers and scientists stepped up their opposition to both projects on ecological and cultural grounds. Their concerns fed into those of Gorbachev, who balked at the high price tags. The government shelved both schemes in August 1986. Now that the union has fractured into sovereign republics, it seems highly unlikely that either project will get off the ground. Ecological and health conditions in the Aral Sea basin, however, are unimaginably bad, and the Siberian transfer scheme remains very much alive in the minds of many central Asians. (See Chapter 5.)[7]

North Americans have their own versions of redistributing northern water wealth toward the drier continental interior. The proposed GRAND (Great Recycling and Northern Development) Canal would block off Canada's James Bay to create a new freshwater reservoir, pump water into the Great Lakes basin, and then channel it from there to the North American west. Devised by Canadian engineer Thomas Kierans, this $100-billion megaproject has the support of Quebec premier Robert Bourassa and, not surprisingly, several powerful engineering firms. Detailed studies of the project still need to be done.[8]

Another scheme, the North American Water and Power Alliance (NAWAPA), first proposed by a U.S. engineering firm in 1964, would transfer a huge volume of water from rivers in western Canada and Alaska to the Canadian prairies, the American southwest, and down into Mexico. Leaving aside the monumental ecological concerns each of these engineering feats raises, neither NAWAPA or GRAND Canal can be economically viable without large government subsidies. The day when farmers and other beneficiaries could profitably pay for the expensive water these projects would deliver is a long way off, and may never arrive.[9]

More recently, Alaska Governor Walter Hickel has proposed shipping some of his state's water wealth to thirsty California by way of an undersea pipeline. The project would involve tapping a southeast Alaskan river, such as the Copper or Stikine, that now delivers large quantities to the sea, and piping its water 2,240 kilometers or more to northern California's Lake Shasta, where it would then enter the state's distribution network.

Ballpark estimates place the cost of the project, which would divert some 5 billion cubic meters per year (4 million acre-feet), at $110 billion. The delivered water would cost $2.40–3.25 per cubic meter ($3,000–4,000 per acre-foot). A study by the U.S. Congressional Office of Technology Assessment concluded that "it does not appear that pipeline water [from Alaska] will ever be able to compete with the more easily implemented supply-enhancing and demand-reducing options now being planned." Besides the unattractive economics, the project poses serious environmental risks, including, for instance, possible temperature and salinity changes in coastal waters that could affect marine life and migrating salmon.[10]

Plans for at least one massive river diversion do seem to be proceeding, though at a slow pace. The Chinese are looking to a large-scale transfer from the Yangtze River in the central part of the country to the Yellow River in order to ease shortages in the water-poor north plain. In February 1983, the government approved the first stage of work on what is known as the East Route, which mainly involves reconstructing the old Grand Canal and so would offer navigation benefits whether or not other phases of the project are completed. Subsequent reports, however, suggest that the diversion scheme is still under review and is only a "possibility." A battery of feasibility studies seems to have turned up numerous obstacles, not least of which is that pumping 20 billion cubic meters of water from the Yangtze up to the higher-elevation Yellow River would require the electrical output of seven large power stations.[11]

The project's initial phase would transfer 5 billion cubic meters of Yangtze water 1,100 kilometers north to Beijing each year. Because of its size, cost, and complexity, the diversion will not be in place soon enough to narrow the troubling water gap projected for Beijing at the end of the decade. Nonetheless, speaking at a press conference there in early 1992, Wu Guocheng, a top water official, said "Diverting water from the Yangtze river to the north is a must if the government hopes to alleviate north China's crippling water shortages. . . . If we don't have the courage and determination to start the diversion project at once, north China's economy will collapse if it is stricken by another extended drought."[12]

Parakrama Bahu would be pleased to learn that not only have we captured a significant share of the world's river flow before it reaches the sea, we have figured out how to extract fresh water from the sea itself. With the oceans holding 97 percent of all the water on earth, desalting seawater would seem to offer an "ultimate solution" to the world's water problems. In 1961, U.S. President John F. Kennedy noted that if humanity could find an inexpensive way to get fresh water from the oceans, that achievement "would really dwarf any other scientific accomplishments."[13]

Thirty years later, the good news is that desalination is technically feasible and use of the process has grown enormously. More than 7,500 desalting plants of various kinds and sizes now operate worldwide, collectively turning 4.8 billion cubic meters of salty water into fresh water each year. The sobering news, however, is that despite its rapid growth, desalination still produces just one tenth of 1 percent of the world's total water use— and its contribution to global water supplies is likely to remain small for the foreseeable future.[14]

Removing salt from water—either by heating it and condensing the steam (known as distillation), or by filtering it through a membrane (reverse osmosis)—takes a great deal of energy. Much early enthusiasm for desalination hinged on nuclear power's promise of energy "too cheap to meter," a hope that never materialized. Today, desalination ranks among the most expensive water supply options. At $1–2 per cubic meter, turning ocean water into drinking water is four to eight times more expensive than the average cost of urban water supplies today—and at least 10–20 times what farmers currently pay.[15]

As a result, desalination remains a solution of last resort. The frequency with which cities and communities are turning to it is more a sign of water scarcity and stress than it is a source of comfort. About 60 percent of the world's desalination capacity is found in the water-poor but energy-rich nations of the Persian Gulf. Saudi Arabia alone claims 30 percent of the global total, with plants scattered along its coastlines distributing desalted drinking water to cities and villages via a 3,000-kilometer pipeline. Island nations, such as those in the Caribbean, and other arid regions, such as Australia and Spain, account for much of the remainder.[16]

Californians, now in their sixth year of drought, are contemplating a major expansion of desalination. As of April 1991, about a dozen plants were under consideration along the coast, including ones in San Diego County, San Luis Obispo, and Marin County. To guard against future dry spells, Santa Barbara has just completed the largest ocean desalting plant in the United States, with a capacity to produce 25,350 cubic meters per day, enough to meet the needs of some 15,000 households. The desalted water will cost $1.57 per cubic meter, four times Santa Barbara's average cost of water prior to the drought. Ironically, right after inaugurating the plant in April 1992, the city mothballed it, since reservoirs were brimming from 1992's heavy spring rains.[17]

Desalting brackish water—which is too salty to drink but much less salty than ocean water—is among the most rapidly growing uses of desalination. At 40–70¢ per cubic meter, it typically costs less than half as much as seawater desalination. Florida communities that have somewhat-salty groundwater have turned to desalination for drinkable supplies. There are now more than

100 mostly small reverse-osmosis plants in Florida alone.[18]

Desalination will be an expensive life saver to a growing number of coastal cities and towns bumping up against supply limits, but it does not constitute the oasis needed in the global water picture. Its costs are way out of line with what farmers, the world's biggest water users, can pay. As with the megadiversion projects that engineers dream of, desalination holds out the unrealistic hope of a supply-side solution, which delays the onset of the water efficiency revolution so urgently needed.

4

Bread
and Water

Living, as so many of us do, in cities, suburbs, and towns, we leave concerns about food production to the farmers and irrigation problems to the engineers. In an age of space travel, telecommunications, and high-tech health care, it seems anachronistic to worry about something as simple as having enough water to grow sufficient food for the world's people.

But there is cause for concern. Water scarcity and increasing competition for limited supplies, poor irrigation practices that damage fertile cropland, and the rising social and environmental costs of large water projects raise doubts about our ability to grow enough food in the years ahead. A close look at water's role and performance in food production reveals the need to build a more stable agricultural foundation.

Crops get the moisture they need to grow in two basic ways—from rainfall and from irrigation. Today, 84 percent of the world's cropland is watered only by rain, while 16 percent benefits from the greater control afforded by farmers applying water. Irrigation has turned many of the earth's sunniest, warmest, and most fertile lands into important crop-producing regions. Egypt could grow very little food without water drawn from the Nile. California's Central Valley, a lucrative fruit and vegetable basket, could barely be cultivated without water delivered by groundwater pumps and surface canals. And absent irrigation, yields in the critical grain-growing areas of northern China, northwest India, and the western U.S. Great Plains would drop by one third to a half.[1]

Some 235 million hectares of land is currently irrigated, a fivefold increase since 1900. (See Table 4–1.) As world population grew from 1.6 billion to more than 5 billion during this century, irrigation became a cornerstone of global food security. A reliable water supply allowed farmers to take advantage of fertilizers and high-yielding seeds to boost their crop output, helping feed the millions of people added to the planet each year. Today, 36 percent of the global harvest comes from the 16 percent of the world's cropland that is irrigated. Many countries—including China, Egypt, India, Indonesia, Israel, Japan, North and South Korea, Pakistan, and Peru—rely on such land for more than half their domestic food production.[2]

For most of modern history, the world's irrigated area grew faster than population did. As a result, the amount of food available per person increased, and, where it was

TABLE 4-1. *Net Irrigated Area, Top 20 Countries and World, 1989*

Country	Net Irrigated Area[1]	Share of Cropland That is Irrigated
	(thousand hectares)	(percent)
China	45,349	47
India	43,039	25
Soviet Union	21,064	9
United States	20,162	11
Pakistan	16,220	78
Indonesia	7,550	36
Iran	5,750	39
Mexico	5,150	21
Thailand	4,230	19
Romania	3,450	33
Spain	3,360	17
Italy	3,100	26
Japan	2,868	62
Bangladesh	2,738	29
Brazil	2,700	3
Afghanistan	2,660	33
Egypt	2,585	100
Iraq	2,550	47
Turkey	2,220	8
Sudan	1,890	15
Other	36,664	7
World	235,299	16

[1]Area actually irrigated; does not take into account double cropping.

SOURCE: See endnote 2.

distributed fairly equitably, diets improved. In 1978, however, per capita irrigated land peaked at 48 hectares for every thousand people, and it has fallen nearly 6 percent since then. (See Figure 4–1.) An estimated

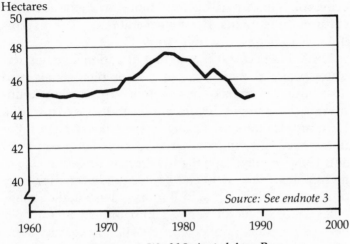

Hectares

FIGURE 4-1. *World Irrigated Area Per 1,000 People, 1961–89*

23.6 million hectares were added to the global irrigation base between 1980 and 1989, based on U.N. Food and Agriculture Organization data—an average of 2.6 million hectares a year, three fifths the rate during the seventies.[3]

This trend of declining irrigated land per person is historically new, and political leaders and development specialists have not yet fully grasped its consequences. Indeed, many have failed even to notice it. Per capita grain production has been falling roughly 1 percent a year since 1984, with the drop concentrated in developing countries. Given that the irrigation slowdown is almost certainly one of the main reasons, food security will continue to erode in a large portion of the developing world. Many people assume that irrigation will pick up speed again once commodity prices rise or general

economic conditions become more favorable for agricultural investments. But for several reasons this seems
unlikely to happen for quite some time.[4]

For one, the cost of expanding irrigation through new
water projects—building dams, reservoirs, canals, and
distribution channels—has risen greatly in many countries, making such schemes harder to finance. In India,
for example, the cost of large canal schemes (in infla-
tion-corrected terms) more than doubled between 1950
and 1980. Today, capital costs for new irrigation capacity run between $1,500 and $4,000 per hectare for large
projects in China, India, Indonesia, Pakistan, the Philippines, and Thailand. They climb toward $6,000 per
hectare for public projects in Brazil, and toward $10,-
000 per hectare in Mexico. In Africa, where roads and
other infrastructure are often lacking and the parcels
that can be irrigated are relatively small, per-hectare
costs have climbed to $10,000–20,000, and sometimes
even higher. Not even double-cropping of higher-valued
crops can make irrigation systems at the top end of this
spectrum economical.[5]

Partially as a result of these high costs, lending for
irrigation by the World Bank and the other major international donors has declined sharply during the last decade. Since large projects take a decade or more to complete, this funding drop suggests that in a good deal of
the Third World, the spread of irrigation will not
quicken much in the next decade.[6]

Meanwhile, many existing systems are functioning
poorly. Over time, distribution channels fill with silt,
outlets break or are bypassed, and infrastructure generally falls into disrepair. Worldwide, an estimated 150
million hectares—nearly two thirds of the world's total

irrigated area—needs some form of upgrading to remain in good working order.[7]

Each year some irrigated land comes out of production entirely as a result of waterlogging and salting of the soil brought about by poor water management. Without adequate drainage, seepage from unlined canals and overwatering of fields raise the level of the underlying groundwater. Eventually, the root zone becomes waterlogged, robbing plants of oxygen and curtailing their growth. In dry climates, evaporation of water near the soil surface leads to a steady accumulation of salt that also reduces crop yields and, if not corrected, can ruin the land. Aerial views of some abandoned irrigated areas show vast expanses of glistening white salt, land so destroyed that it is essentially useless.[8]

No one knows for sure how large an area suffers from salinization. International irrigation consultant W. Robert Rangeley estimates that 15 million hectares in developing countries—primarily China, India, Iran, Iraq, and Pakistan—experience serious reductions in crop yields because of salt buildup in the soil. Studies by the World Bank have found that waterlogging and salinity are reducing yields of major crops in Egypt and Pakistan by 30 percent. And in Mexico, salinization is estimated to be curtailing the annual grain harvest by the equivalent of 1 million tons—enough to feed 5 million people, more than a quarter of Mexico City.[9]

In the United States, salinity expert James Rhoades estimates that salt accumulation is lowering crop yields on 25–30 percent of the nation's irrigated land, or more than 5 million hectares. Some 2.5 million hectares are salinized in the former Soviet Union, most of them in the irrigated deserts of central Asia. All told, some 25

million hectares—more than 10 percent of world irrigated area—appears to suffer from yield-suppressing
salt buildup. And the problem is growing: salinization is
spreading at an estimated rate of 1–1.5 million hectares
each year, about half the rate at which new land is being
brought under irrigation.[10]

Another sign of trouble is the amount of food currently produced by overpumping groundwater. In the
United States, more than 4 million hectares—roughly a
fifth of the nation's irrigated area—is watered by pumping in excess of recharge. By the early eighties, the depletion was particularly severe in California, Kansas, Ne-
braska, and Texas, four important food-producing
states. By definition, overpumping cannot continue in-
definitely. If farmers do not bring their water use into
balance with natural recharge, eventually the water will
become too costly to pump to the surface or will run out
altogether, and the land will be forced out of irrigation.[11]

No other country has systematically assessed the extent of excessive groundwater pumping. But the situation is serious elsewhere as well, including in China and
India, two of the three other major food producers.
Groundwater levels are falling by up to a meter per year
in parts of northern China, and heavy pumping in portions of the southern Indian state of Tamil Nadu reportedly dropped water levels as much as 25–30 meters in a
decade. In the western state of Gujarat, overpumping by
irrigators in the coastal districts has caused saltwater to
invade the regional aquifer, contaminating village drinking supplies.[12]

Although large-scale irrigation has buttressed the
world against famine and helped eliminate some pockets
of chronic hunger, it often has not served other important development goals, such as reducing poverty, pro-

moting equity, protecting natural systems, and improving human health. As a result, new schemes are coming under careful scrutiny. Many people are beginning to ask, Irrigation for whom, and at what social and environmental cost?

India, with nearly one fifth of the world's irrigated land, illustrates the need for a rethinking of ends and means. The nation seems to be in an all-out race to harness its entire irrigation potential (estimated at 113 million hectares, or 2.6 times the present area), even though much of the existing irrigated area is not in use or is performing poorly. Prime Minister Rajiv Gandhi bluntly criticized the nation's track record in a 1986 speech to state irrigation ministers. Of the 246 large surface-irrigation projects started since 1951, only 65 had been completed by then, he said, and little benefit had come from projects begun after 1970. "For 16 years we have poured money out. The people have got nothing back, no irrigation, no water, no increase in production, no help in their daily life."[13]

Today, the fierce battle being waged over the Narmada Valley Development Program in western India encapsulates this heightened questioning of irrigation's proper role. Among the largest water development ventures under way in the world, it encompasses 30 large dams, 135 medium-sized ones, and 3,000 small ones. The reservoir to be created by the Sardar Sarovar dam, the scheme's centerpiece, would alone flood 37,000 hectares of forest and farmland and displace up to 100,000 people, mostly poor tribal villagers.[14]

Although about 90 percent of the lands to be submerged and people to be relocated are in the states of Madhya Pradesh and Maharashtra, the bulk of the project's estimated benefits—including irrigation water for

1.8 million hectares—will go to Gujarat. Few dispute
that Gujarat needs more water, but concerns have been
raised that waterlogging and salinization in the newly
irrigated area could render the Narmada scheme's benefits short-lived. Opponents have persistently claimed
that the project will not serve the most destitute areas or
people. They point out, for instance, that nearly 70 percent of Gujarat's drought-prone areas and 90 percent of
tribal areas will receive no water. As Baba Amte writes in
Cry, the Beloved Narmada, Sardar Sarovar "will draw
money away from various other schemes which could
provide water to these areas [in need]. . . . The government has completely lost track of what must be regarded
as its basic objective: finding the best possible way of
providing water to the people."[15]

Strong opposition to Sardar Sarovar from villagers in
the Narmada valley, as well as from environmentalists
and human rights advocates in India and elsewhere, has
forced a reevaluation of the project's aims and consequences. The discontent stems not just from the
scheme's apparent inequities but from insufficient attention to its environmental impacts and concern about
the resettling of those whose homes will be submerged.[16]

An independent review commissioned by the World
Bank, which has approved $450 million for Sardar Saro-
var (about 8 percent of the base cost), reported in June
1992 that the project had never been properly assessed,
and that serious weaknesses exist in the resettlement
plans and in environmental protection measures. It concluded that "it would be prudent if the necessary studies
were done and the data made available for informed
decision-making before further construction takes
place. Implementation requires that the Bank take a step
back." The reviewers' findings will almost certainly

lessen other sources of international support for the project as well.[17]

Large irrigation projects have also compromised human health by contributing to the spread of debilitating waterborne diseases. Among the worst of them is bilharzia, caused by a parasitic worm released by aquatic snails, which then penetrates human skin that comes in contact with the infected water. Also known as schistosomiasis, bilharzia causes blood loss and disorders of the liver, bladder, lungs, and nervous system. It is widespread in parts of Africa, the Middle East, and South America, and has appeared in China and Southeast Asia as well.[18]

According to agricultural economist Jose Olivares of the World Bank, in Africa "the scale of disease connected with irrigation is massive. . . . Horror stories are told in almost every country." Among the most extreme cases he cites is the Gezira project in Sudan, where the prevalence rate for bilharzia ranged from 5–10 percent of the population at risk before the irrigation scheme came on-line and then skyrocketed to more than 80 percent after its completion. Weakened and debilitated by this terrible disease, the project's "beneficiaries" have almost certainly lost in terms of health more than they have gained from greater access to irrigation water.[19]

Finally, as discussed in later chapters, increased competition for scarce water along with a host of environmental concerns—from toxic irrigation drainage in the American West to the shrinking Aral Sea in central Asia—will siphon supplies away from agriculture. Thus not only will new irrigation projects be harder to build, but some existing irrigated lands will lose water in order to restore degraded ecosystems and to deliver drinking water to growing cities.

With some 95 million people being added to the planet each year in the nineties, new strategies will be needed to prevent the many emerging irrigation constraints from leading to food shortages. It now seems likely that irrigated area will spread by no more than 1 percent a year for the foreseeable future, while world population expands by 1.7 percent annually. Turning to desalination or long-distance water transfers does not offer a practical solution, since the food produced with such expensive water will be out of reach for those most at risk—the 1 billion people now subsisting on less than $1 a day.[20]

New technologies will relieve some of the constraints now emerging. Breeding and selecting crop varieties that are more salt-tolerant, drought-resistant, and water-efficient will help bolster crop production as freshwater supplies for agriculture dwindle. Studies suggest, for instance, that wheat is a good candidate for breeding in greater tolerance to salt, which could allow this important grain to remain productive on salinized land. Researchers are also learning how to better match crops to varying qualities of water. Israelis are irrigating cotton, corn, tomatoes, and asparagas with water more than twice as salty as the limit recommended for drinking in the United States.[21]

The biggest gains for the forseeable future, however, will come from irrigating crops more efficiently. As described in Part II, this can free up supplies to expand irrigated land in nearby areas while avoiding the high cost and environmental harm posed by new water development projects. In addition, better water management can boost production on existing irrigated lands that are now yielding well below their potential, including an estimated 10–13 million hectares in India alone.[22]

Where water supplies do need to be expanded, smaller-scale projects—including, for example, shallow groundwater wells, garden irrigation, and small reservoirs for storing local runoff—show much potential for increasing food production cost-effectively and with less damage to the environment. As described in Chapter 9, many of these schemes can be funded and developed privately by local people if they have access to appropriate technologies and credit to purchase them. And, finally, devoting far more attention and resources to raising crop output on the 84 percent of the world's cropland watered only by rainfall can do a great deal for global food security, as well as for the income security of many of the world's poorest farmers.

5

Paradise
Lost

Hanging on the office wall of Soviet parliamentarian
Alexei Yablokov a few years ago was a map with a glar-
ing omission. The cartographers who prepared it in the
early sixties depicted a future without the Aral Sea, then
the world's fourth largest freshwater lake. Their geographic rendering represents what is surely one of the
heights of human arrogance: the planned elimination of
an ecosystem nearly the size of Ireland.[1]

Today, the decimation of the Aral Sea in central Asia
ranks among the more dramatic in a long list of natural
areas destroyed, degraded, or at grave risk from human
use and abuse of water. The damming, diverting, and
polluting of watercourses with little regard for the environmental services they provide and the species they
support has wreaked havoc on the world's wetlands, del-

tas, lakes, and riverine habitats. Of all threatened forms of biological diversity on earth, aquatic life may be the most in jeopardy.

A distressing conflict has emerged over two of water's roles: as a commodity serving the economic aims of greater agricultural productivity, industrial expansion, and urban growth, and as a key life-support for all species and natural communities. Mounting scarcity has thrown this friction into sharp relief. More water devoted to human needs means less for sustenance of ecosystems—and, in many areas, nature is losing out fast.

With good reason, the shrinking Aral Sea has come to symbolize much of what has gone wrong with water management. Soviet central planners mapped out its demise based on a simple calculation that using central Asian rivers for irrigation would produce more economic value than letting them flow unimpeded into the Aral. Irrigated area in the Aral Sea basin expanded by half during the past three decades, reaching 7.5 million hectares and creating a lucrative area of farmland. Before the breakup of the Soviet Union, the region produced a third of the nation's fruits and grapes, a quarter of its vegetables, 40 percent of the rice, and 90 percent of the cotton. Irrigation water came primarily from the Amu Dar'ya and Syr Dar'ya (dar'ya means river in Turkic), which, along with meager rainfall, are the only sources of replenishment for the Aral Sea. By 1980, agricultural water consumption had reduced flows in these rivers' lower stretches to a trickle.[2]

Cutting off the Aral Sea's lifelines has had disastrous effects. Since 1960, the sea's surface area has shrunk more than 40 percent, volume has dropped more than 60 percent, and salinity levels have tripled. Each year, winds pick up at least 40 million tons of a toxic dust-salt

mixture from the dry seabed, and dump them on surrounding croplands. According to V. M. Kotlyakov, as
director of the Institute of Geography at the Soviet
Academy of Sciences, dust storms "so powerful they
can be observed from space have become common in
spring." All 24 native fish species in the Aral have disap-
peared, and the fish catch, which totaled 44,000 tons in
the fifties and supported 60,000 jobs, has dropped to
zero. Abandoned fishing villages now dot the sea's former coastline.[3]

Tragically, residents remaining in the Aral Sea basin
offer living proof that loss of human health and well-
being follows close on the heels of environmental decline. Low river flows have concentrated salts and toxic
chemicals, making water supplies hazardous to drink.
Coupled with poor sanitary conditions and heavy pesti-
cide use, contaminated drinking water has contributed
to rampant disease. The incidence of typhoid fever has
risen nearly thirtyfold, and that of hepatitis, sevenfold.
The rate of esophageal cancer in Muynak, an old fishing
port, is 15 times the Soviet average.[4]

Despite President Gorbachev's shelving of the Siberian river diversion scheme in 1986, central Asian political leaders maintained that "solving" the Aral Sea crisis
was impossible without diverting a portion of Siberian
river flows into the region, and they repeatedly called on
Moscow for help. With their newly won independence,
the Asian republics no longer have such recourse. The
fate of the Aral Sea now rests with the six republics in its
basin, and depends on whether, once the political dust
settles, cooperation or conflict emerges from their common predicament. Some help could come from Russia,
which may want to protect the agricultural potential of
its Commonwealth partners. Certainly, the herculean

task of returning the region to ecological health will take a heavy dose of international support.[5]

Botswana's Okavango Delta is ripe for a similar tug-of-war between the water demands of conventional economic development and those of a healthy ecosystem. The government of this southern African nation has proposed siphoning off some of the delta's water to expand irrigation by 1 million hectares, provide drinking water to the northern city of Maun, and increase supplies to the Orapa diamond mine. The plan would involve dredging, damming, and diverting the Boro River that emerges from the delta, a watery wilderness formed where the Okavango River fans out after crossing the border with Angola. Africa's largest oasis, the Okavango Delta is world-renowned for its diverse wildlife, which includes zebras, antelope, elephants, and Cape buffalo. During the dry season, animals migrate from the nearby Kalahari Desert and the wildlife population supported by the delta climbs some tenfold.[6]

At least for the moment, the diversion project is stalled. When the government invited local people to voice their views on the plan in early 1991, they resoundingly expressed opposition. Among the most vocal were herders and fishers, whose livelihoods depend directly on the integrity of the delta. *Washington Post* correspondant Neil Henry quotes one fisher who expressed his reverence for the Boro at a public meeting with government officials: "We believe this river has a life of its own. . . . It is not for man to kill it."[7]

Allied with western conservationists and the profitable Botswana tourist industry, those subsisting on the delta's natural wealth appear to have won out, at least for the time being. In mid-1992, after release of a report by the World Conservation Union showing less need for

the water supply scheme than originally thought, the
government decided to cancel the project. But with rich
cattle barons and a lucrative diamond mine standing to
benefit from more water diversions, the Okavango oasis
could again come under the threat of water exploita-
tion.[8]

Similar conflicts exist in many other African countries
desiring to enlist water's aid in economic development.
The Okavango Delta is but one of many shallow river
basin plains scattered across the continent. About 25
wetland areas of more than 100,000 hectares exist, together encompassing some 30 million hectares. A mix of
swamps, shallow lakes, and floodplains, these areas offer
a multitude of benefits, including flood control and hydrological stability, rich harvests of fish, subsistence
livelihoods for pastoralists and other rural people, and
critical habitat for migratory birds and other wildlife.
They also "consume" large amounts of water through
evaporation, water that might otherwise be available to
irrigate more land or supply more city-dwellers. This
gives rise to the desire to drain or divert some of these
waters before they are "lost"—a process that inevitably
damages the ecosystem.[9]

The Sudd wetlands on the White Nile in southern
Sudan—which, at 9.2 million hectares, form one of the
largest swamps in the world—are slated for just such a
project. Sudan and its downstream neighbor, Egypt,
have jointly planned several projects to increase the
amount of Nile water available for irrigation and overall
economic development. One, called the Jonglei Project,
would involve constructing a large canal to channel
water through the Sudd swamps so as to reduce huge
evaporation losses—estimated at 34 billion cubic meters
per year. The initial phase of the Jonglei, which would

capture an additional 4 billion cubic meters, was 70 percent complete when the Sudanese civil war halted work in 1983. A second phase would bring the increase in Nile supplies to 7 billion cubic meters. Although the project has been stalled for nearly a decade and is opposed by the Sudanese People's Liberation Army, it remains a priority of both the Sudanese and Egyptian governments.[10]

A treasure trove of wildlife, the Sudd swamps are home to millions of migratory birds during the course of a year, including many varieties of storks, cranes, ibis, and herons, to name a few. The largest single bird population, the glossy ibis, reaches some 1.7 million during the early dry season. The Jonglei Canal and other projects designed to reduce evaporation from the Sudd wetlands will inevitably shrink the habitat available for these species, while augmenting the water supply for food production and economic development in Egypt and Sudan.[11]

As with the Soviets' early calculations regarding water use in the central Asian republics, Egypt and Sudan judge water's value for irrigated agriculture to exceed that of ecosystem protection in the Sudd region. That equation could shift, however, if wealthier countries desiring to protect unique wildlife habitat were to offer sufficient compensation to Egypt and Sudan to leave the Sudd wetlands intact. Rough calculations by Dale Whittington and Elizabeth McClelland of the University of North Carolina suggest that a lump sum of $5 billion would match the present value of additional irrigation water from all the upper–White Nile projects now planned, including the Jonglei Canal.[12]

Across the Atlantic, another cherished wetland wilderness has taught us volumes about nature's limits—

and the consequences of overstepping them. South
Florida's Everglades is buckling under pressure from
pollution and water diversions to meet the demands of
agriculture and a rapidly growing population. According
to Robert Arnsberger, assistant superintendent of Everglades National Park, the stressed-out system "could
ecologically fail within the next 20 years."[13]

Known as "grassy water" to the region's native Americans, the Everglades suffered extensive damage as engineers built dikes, canals, levees, and pump stations to
control flooding and siphon off water for farmers,
coastal residents, and tourists. Only half the original 1.6
million hectares of wetland remain, about 200,000 of
them within the boundaries of Everglades National
Park. The disruption of seasonal water patterns and
dwindling of habitat have caused populations of nesting
wading birds—including herons, egrets, ibis, and wood
storks—to plummet from some 300,000 in the thirties
to no more than 30,000 today. Fourteen endangered
wildlife species are hanging on in the park, including the
wood stork, the American crocodile, and the Florida
panther.[14]

Pollution jeopardizes the Everglades' future as well.
Just south of Lake Okeechobee, some 280,000 hectares
of agricultural land—nearly two thirds of it in sugar
cane—not only claims the lion's share of south Florida's
water but also releases runoff contaminated with nitrogen and phosphorus into adjacent wetlands. Elevated
levels of these nutrients foster the growth of cattails,
which have clogged waterways and caused shifts in the
natural vegetation. Wetlands filter out most of the excess nutrients before they reach the park, but continued
heavy loads could cause a pollution front to move south,
threatening the park's ecological health.[15]

In an unusual twist, the federal government filed suit in 1988 against the regional management agency overseeing the water system and other Florida officials, claiming that lax enforcement of state water quality standards placed the Everglades Park and the Loxahatchee National Wildlife Refuge at risk. In July 1991, the parties reached a settlement that calls for converting at least 14,000 hectares of farmland into artificial wetlands that will trap a good portion of the agricultural nutrients. A new tax district has been authorized within the agricultural area in order to help pay for an Everglades rescue effort. For its part, the U.S. Congress appropriated funds in 1990 for a 43,000-hectare expansion of the park's eastern boundary and asked the U.S. Army Corps of Engineers (which, ironically, had built most of the water system) to develop a plan for recreating original water patterns there.[16]

Although these measures inspire hope, they by no means guarantee that the Everglades is safe. Funding for the agricultural cleanup effort, estimated at $400–600 million, is not yet secure. The sugarcane growers and other agricultural interests have filed petitions for administrative hearings on the plan, and have submitted an alternative strategy, which is under study. And with a large influx of people to Florida each year and with each resident using on average 760 liters of water per day— well above the national average—serious urban conservation efforts must also be part of the solution. Ultimately, urban growth must slow if the Everglades is to have a fighting chance. If such a paradise is lost in one of the wealthiest countries in the world, what realistic hope is there for wild places elsewhere?[17]

More widespread trouble is brewing in the western United States, particularly in California. With more people than in all of Canada, the largest irrigation project in the western hemisphere, and a highly skewed distribution of rainfall, California has placed great stress on its natural water system. The federally built $2.1-billion Central Valley Project and California's own State Water Project together create an elaborate plumbing system of pumps, dams, reservoirs, and aqueducts that has given the state one of the richest agricultural valleys in the world and an urban landscape teeming with swimming pools, car washes, and green lawns.[18]

Not surprisingly, this massive water development has decimated California's aquatic ecosystems and wildlife populations, a situation compounded by the ongoing drought. The Sacramento River's winter run of chinook salmon declined from 120,000 in the sixties to just 400 today, and the species was added to the federal endangered list in 1989. The tiny delta smelt, which thrives where ocean water meets fresh in the Sacramento-San Joaquin delta northeast of San Francisco, has seen its numbers drop 90 percent over the last two decades, as freshwater diversions lifted salinity levels beyond the smelt's tolerance limit and as huge numbers of fish got sucked into powerful water pumps. Designation of the unassuming fish as endangered could force cutbacks in the volume of water siphoned away from the delta. So far, officials have only proposed listing the fish as "threatened," a designation biologists believe may not be sufficient to save the smelt from extinction.[19]

Western wetlands and wildlife have also been ravaged by toxic chemicals in agricultural drainage water—in particular, selenium, a natural element essential to humans and other life but poisonous at elevated concentrations. Irrigation has washed large quantities of selenium out of the soil, with the resulting toxic drainage often ending up at wetlands and wildlife refuges. Alarming discoveries of death, grotesque deformities, and reproductive failure in birds and other animals at the Kesterson National Wildlife Refuge in California's Central Valley sparked initial concerns in 1983, and the problem has since grown to staggering proportions.[20]

Scientists have uncovered patterns of death and deformity similar to that in Kesterson at five other U.S. sites: the Tulare Basin, also in California's Central Valley; the Stillwater National Wildlife Refuge in Nevada, a haven for the tundra swan, avocet, bald eagle, and more than 160 other bird species; the Ouray National Wildlife Refuge in northeastern Utah; the Kendrick Project near Casper, Wyoming; and the Benton Lake National Wildlife Refuge in Montana. At six others, selenium levels exceed toxicity thresholds, but researchers have not yet confirmed harm to wildlife. Investigations have been recommended or are under way at 18 other locations.[21]

Cleaning up each site will be expensive. Proposed management plans for California's San Joaquin valley alone would cost some $40 million per year, and exactly what constitutes site-cleanup remains unclear. The federal government ordered Kesterson to be dried out, which has left behind toxic vegetation and the likely possibility that selenium continues to accumulate in the food chain. Writes environmental journalist Tom Harris in *Death in the Marsh*, a detailed account of the ongoing selenium drama, it looks like "dead and deformed rabbits, songbirds, rodents, and hawks will soon retrace the tragic destiny of coots, ducks, frogs, and fish." Neither the federal government nor the western states have come to grips with the full extent and severity of irrigation's toxic legacy, and no adequate plan for defusing the threat yet exists.[22]

Another sign of the severely compromised health of aquatic ecosystems is the 364 species of fish in North America listed by the American Fisheries Society as endangered, threatened, or of special concern—the vast majority of them at risk because of habitat destruction. Throughout Canada, the United States, and Mexico, an estimated one third of the fish, two thirds of the crayfish, and nearly three fourths of the mussels are now "rare or imperiled."[23]

They often come to such status by way of incremental human actions that end up undermining their basic habitat requirements—be it the timing, quantity, or quality of water's flow. In the Mexican state of Nuevo León, for instance, three endemic species (those found nowhere else) live in a shallow marshland and reservoir at the Ojo del Potosi, where several springs emerge from the base of a cliff. In 1985, intensified groundwater pumping in the region shrunk this habitat to only 15 percent of its former size. By 1987, it was down to only 5 percent, and by 1989, only a shallow irrigation ditch was left. Some of the stocks were removed to be maintained in a laboratory, but for two of the species, propagation has proved difficult.[24]

Similar tales could be told for many Mexican fish species on the threatened list. As Salvador Contreras Balderas, a scientist at the Universidad Autonoma de Nuevo Leon, says, "the situation for aquatic habitats and freshwater fishes in Mexico is not promising . . . an appalling number of endemic species and affected aquatic habitats will almost certainly disappear before an adequate and rational system of species conservation and protected areas can be attained."[25]

Of the many varieties of native fish species at risk in North America, perhaps the most notable for their cultural and recreational values are several species of salmon in the western United States. In 1991, just four adult Snake River sockeye salmon made it from the Pacific Ocean past eight federal dams in the Columbia River basin to their primordial spawning ground at Idaho's Redfish Lake. On the brink of extinction, the Snake River sockeye was listed as endangered in November 1991. And as of early July 1992, steps were being taken to move the Snake River chinook salmon from threatened to endangered status.[26]

The U.S. Endangered Species Act offers a powerful lever for protection of aquatic ecosystems, since it can force human activity to adjust to the habitat needs of a species found to be at risk of extinction. In the case of western salmon fisheries, this could require dam operators to partially recreate natural river flow patterns by releasing more water each spring, when baby salmon make their drive toward the sea. Typically, spring runoff is held back in giant reservoirs for release later in summer, when shipping, farming, and hydropower production require more water. Salmon fry end up dying in huge numbers for lack of sufficient current to aid their seaward journey. Boosting their survival rate could entail significant cutbacks in hydropower and irrigation, setting the stage for an endangered species controversy that may rival the spotted owl dispute that has altered western timber harvests.[27]

At the heart of the conflict between water's use for economic gain and for the maintenance of aquatic ecosystems is the extent to which private rights to water may be circumscribed to protect common public claims to water's environmental services. In the United States, legal guidance on this question dates back to early this century, when Justice Oliver Wendell Holmes, in upholding New Jersey's right to prohibit a water company from diverting a portion of the Passaic River, said that "the private right to appropriate is subject not only to the rights of lower owners but to the initial limitation that it may not substantially diminish one of the great foundations of public welfare and health."[28]

Unfortunately, only in rare instances are public values and future generations winning out over private rights to dam and divert natural watercourses. And in the United States, a growing movement to protect property rights from government actions to safeguard the environment could tip this balance even further away from ecosystem protection.

Each threatened wetland, lake, or aquatic species presents a crucial test of whether a region's people and economy can adapt to the ecological needs of a healthy aquatic system. Few water treasures in poor nations stand a chance without international help. In the end, a scant number will survive anywhere unless we reclaim some respect for water systems and the life they support, recognize that the health and diversity of aquatic species are key indicators of the overall well-being of the environment, and begin to accept that our own destiny is inextricably linked to that of the water world around us.

6

Hydropolitics

In a 1989 address before the U.S. Congress, Boutros Boutros-Ghali, then Egypt's Minister of State for Foreign Affairs, noted: "The national security of Egypt is in the hands of the eight other African countries in the Nile basin."[1]

Besides underscoring water's importance to Egypt's economy, Boutros-Ghali's statement reflects one of water's unique attributes. Not only does it course easily across political boundaries, it gives upstream countries a distinct advantage over downstream neighbors. As population pressures and rising demands press against the limits of supplies, international frictions over water are intensifying.

Nearly 40 percent of the world's people live in river basins shared by more than two countries. India and

Bangladesh haggle over the Ganges River, Mexico and the United States over the Colorado, Czechoslovakia and Hungary over the Danube, and Thailand and Vietnam over the Mekong. Africa alone contains 57 river and lake basins shared by at least two nations. Nowhere, however, are water disputes shaping political landscapes and economic futures as definitively as in the Middle East.[2]

Talk of a "water crisis" in the Middle East has become almost legendary. With some of the highest population growth rates in the world and heavy reliance on irrigation for their agricultural productivity, Middle East countries have much at stake when it comes to distributing the region's supplies. Enough leaders have spoken of the potential for wars over water that new warnings have begun to lack bite. But far from crying wolf, these repeated admonitions presage momentous events in Middle East politics. Over the next decade, water issues in the region's three major river basins—the Jordan, the Nile, and the Tigris-Euphrates—will foster either an unprecedented degree of cooperation or a combustible level of conflict.

Israel, Jordan, and the occupied West Bank share the waters of the Jordan River basin. (See Figure 6–1.) Israel's annual water use already exceeds its renewable supply by some 300 million cubic meters, or 15 percent. With the projected influx of up to 1 million Soviet Jews over the next decade, its yearly water deficit will worsen greatly. Jordan's water use, less than half Israel's on a per capita basis, is also bumping up against supply limits, and its population grows by 3.4 percent annually, among the highest rates in the world. With the nation's water demand projected to increase by 40 percent during this decade, competition grows keener each year.

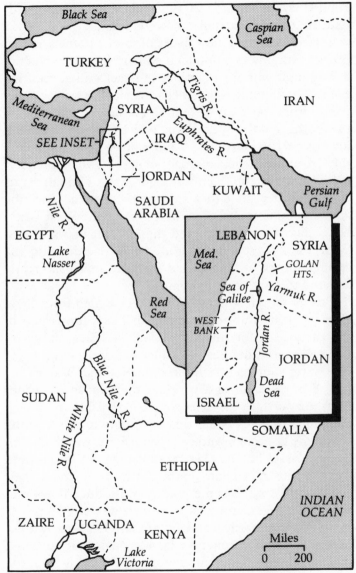

FIGURE 6-1. *Middle East River Basins*

King Hussein declared in 1990 that water was the only issue that could take him to war with Israel.[3]

Solutions to the Palestinian quest for a homeland that propose trading Israeli-occupied land for peace will not be workable without assurance of Israel's water security. Some 25–40 percent of Israel's sustainable water supply comes from the Yarqon-Taninim aquifer running along the foothills of the West Bank, which Israel claimed following the 1967 war. This aquifer flows westward across the Green Line (the demarcation of pre-1967 Israeli territory) toward the Mediterranean Sea. Though Israel can tap water on either side of the Green Line, the aquifer's main recharge area lies under the West Bank. Israel has severely restricted the amount of water that West Bank Arabs can pump, while overdrawing the aquifer for its own uses—an inequity that has greatly angered the Arab population.[4]

The Golan Heights, which Israel claimed from Syria after the 1967 war and annexed in 1981, forms part of the catchment for the Sea of Galilee. Also known as Lake Kinneret, the Sea of Galilee is Israel's largest surface water source and supplies the National Water Carrier, a huge canal and pipeline that transports water from the north to the drier south. Control of the Golan Heights also gives Israel riparian rights to the Yarmuk River, the last major undeveloped tributary in the basin. So far, Israel has blocked a joint plan by Jordan and Syria to construct a dam on the Yarmuk to increase their supplies, fearing that the dam could reduce flows into the Jordan River, and thus jeopardize its water security.[5]

An extraordinarily wet winter in late 1991–early 1992 brought welcome relief to Israelis, who had suffered through several consecutive years of drought. Still, the 1992 rains only temporarily brighten a grim picture.

Decades of overpumping have caused seawater to invade Israel's coastal aquifer, a key freshwater source. Some 20 percent of the aquifer is now contaminated by salts or by nitrates from urban and agricultural pollution, and water officials predict that a fifth of the coastal wells may need to be closed over the next few years. This degradation of the coastal aquifer greatly deepens Israeli dependence on the reserve underlying the West Bank. As Thomas Naff, Middle East water analyst at the University of Pennsylvania, sums it up, "It is water, in the final analysis, that will determine the future of the Occupied Territories, and by extension, the issue of conflict or peace in the region."[6]

Conservation and increased efficiency, particularly in agriculture, can help stave off chronic and debilitating shortages in the near term. (See Chapter 8.) But supplies in the region are so tight that it will take a fundamental economic restructuring, including cutbacks in irrigated agriculture, along with population stabilization and basin-wide water management to achieve any degree of water security. A war over water might create some temporary winners. But unless the nations of the Jordan River basin cooperate to find equitable solutions to their common plight of scarce water, lasting peace will remain elusive.

Across the Sinai Peninsula, tensions are high—if less ignitable at the moment—in the Nile River basin as well. In Egypt, 56 million people depend almost entirely on the waters of the Nile, none of which originates within the nation's boundaries. About 85 percent arises in Ethiopia, and flows as the Blue Nile into Sudan. The remainder comes from the White Nile system, with headwaters at Lake Victoria in Tanzania, which joins the Blue Nile near Khartoum. The world's longest river, the Nile supplies nine countries in all—of which Egypt is last in line.[7]

Under a 1959 agreement with Sudan, Egypt is entitled to 55.5 billion cubic meters (BCM) of Nile water per year, while Sudan is allotted 18.5 BCM. To meet its needs, Egypt supplements Nile water with small amounts of groundwater, agricultural drainage water, and treated municipal wastewater; the total amount available from all sources in 1990 was 63.5 BCM. Even modest projections show demand rising to 69.4 BCM by the end of the decade, a 17-percent increase over the 1990 usage level and 9 percent greater than current supplies.[8]

The strategy for meeting this demand depends on some questionable components—including the extraction of deep groundwater in the desert, a fivefold increase in treated sewage water, and increased availability of Nile water from the Jonglei project with Sudan. With construction on this project having stopped in 1983 and the Sudanese People's Liberation Army adamantly opposed to it, this effort to reduce evaporation from Sudan's Sudd swamps is unlikely to yield additional water for Egypt in the foreseeable future, if ever.[9]

Clearly, any loss of Nile water would hasten and exacerbate Egypt's impending water crunch. Drought in the African Horn can substantially reduce supplies, as it did in 1984–85, when the Nile's flow into Egypt dropped to 38 BCM, nearly a third less than Egypt's allotment. Moreover, since upper-basin countries recognize no obligation to limit their use of Nile waters for the sake of Egypt and Sudan, their water development projects could permanently reduce flows downstream. It was such a fear that led Egyptian President Anwar Sadat to state, shortly after signing the historic peace accords with Israel, that "the only matter that could take Egypt to war again is water."[10]

Fortunately for Egypt, Ethiopia's plans to dam upper Nile waters have not come to fruition. Lake Tana, the Blue Nile's source, is remote, greatly complicating development efforts that are already hindered by political turmoil and economic disarray. How long the Blue Nile will flow freely through Ethiopia to the disproportionate benefit of Egypt is, however, uncertain. In early 1990, Egypt was reported to have temporarily blocked an African Development Bank loan to Ethiopia for a project that Cairo feared would reduce downstream supplies.[11]

Solutions to Egypt's predicament would seem to have a recipe similar to that for the Jordan basin: braking population growth, increasing irrigation efficiency and recycling more water, shifting some supplies out of agriculture, and building water security through regional compacts. Unfortunately, meaningful cooperation among countries in the Nile Basin—in particular Ethiopia, Egypt, and Sudan—does not appear likely in the near future. This became clear in June 1990 at a gathering in Cairo dubbed the African Water Summit, at which Nile issues received prominent attention. Several Nile basin nations sent representatives to the high-level meeting, including Ethiopia, an important departure from that nation's past practice.

Little was achieved at the meeting. Ethiopia was unwilling even to share basic hydrological data with its neighbors without a decision to renegotiate the 1959 agreement between Egypt and Sudan. Because that agreement allots so much of the basin's total flow to two downstream countries that contribute practically nothing to that flow, Ethiopia views the compact as unworkable.[12]

A forum for basin-wide cooperation does exist through a group called Undugu, Swahili for "fraternity," in which all the Nile states have participated. Yet until Egypt and Sudan recognize the need to reopen the allocation issue in fairness to Ethiopia, no agreement encompassing all the basin countries seems possible.

Perhaps the best hope for breaking the logjam lies in cooperative solutions that can benefit all parties. Dale Whittington and Elizabeth McClelland of the University of North Carolina have pointed out, for instance, that storing more of the Nile's water in reservoirs in the Ethiopian highlands, where evaporation is much lower than downstream, could result in more water for all three Blue Nile basin countries—Ethiopia, Sudan, and Egypt. According to their analysis, enough water could be saved this way to nearly quadruple Ethiopia's irrigated area without reducing supplies to Egypt and Sudan.[13]

A cooperative agreement would also enable Ethiopia to receive water project funding from the World Bank and other international sources that will not support projects that may harm neighboring countries. Such a scheme, however, would alter the ecology of the Ethiopian Highlands, and would need to be carefully studied. But mutually beneficial approaches of this sort can help defuse the potential for conflict as basin-wide water stresses mount.

Of the three Middle East river basins, only the Tigris-Euphrates has substantial water left after current demands are met. Yet this relative abundance has not prevented tensions from arising, nor will it likely last for long. Here, too, the failure of the basin's three countries—Iraq, Syria, and Turkey—to reach water-sharing agreements has created an atmosphere of competition and mistrust that could breed future conflict.

The mountains of eastern Turkey give rise to both rivers, with the Euphrates flowing through Syria and Iraq before reaching the Persian Gulf and with the Tigris running directly through Iraq to the Gulf. Oil-poor but rich in water and agricultural land, Turkey has undertaken a massive water development scheme called the Southeast Anatolia Project designed to boost its hydropower capacity by some 7,500 megawatts and its irrigated area by half, as well as to promote economic development in the region. Referred to as the GAP, after the Turkish acronym, the Anatolia scheme includes construction of 25 irrigation systems, 22 dams, and 19 hydropower stations.[14]

Syria and Iraq fear that this huge endeavor could foil their own development plans and leave them short of water. The GAP could reduce the Euphrates's flow into Syria by 35 percent in normal years and substantially more in dry years, besides polluting the river with irrigation drainage. Last in line, Iraq also worries about Syria's plans to tap more of the Euphrates for irrigation and to meet the domestic needs of a population that, at current growth rates, will double in 18 years. Damascus, Aleppo, and other Syrian cities have already experienced supply cutbacks in recent years. All three countries in the basin weathered water shortages in 1989 when drought cut the Euphrates's normal flow in half.[15]

In January 1990, Turkey heightened the anxieties of its downstream neighbors by stopping for one month the flow of the Euphrates below the Ataturk Dam, the GAP's centerpiece and now the fifth largest rockfill dam in the world. Turkey had told Syria and Iraq the previous November of its plans to start filling the reservoir behind the dam, and offered to compensate them by increasing downstream flows from November until mid-January. Nonetheless, Syria and Iraq protested Turkey's action. President Turgut Ozal tried to reassure them that Turkey would never use its power over the river to "coerce or threaten them." The assurance rang a bit hollow, however, given his government's veiled threat in late 1989 to cut the Euphrates's flow because of Syria's support of Kurdish insurgents.[16]

Turkey has underscored its role as water broker in the region by proposing to build what it calls "peace pipelines" to drier Middle East nations. A western pipeline would deliver drinking water to cities and towns in Jordan, Syria, and Saudi Arabia; another would follow a Gulf route and take supplies to Kuwait, Saudi Arabia, the United Arab Emirates, Qatar, Oman, and Bahrain. Along with the pipelines' $21-billion estimated cost, a major drawback is that financing would depend on all parties reaching a broader water-sharing agreement. Perhaps a larger hurdle is that at this point the downstream Arab nations do not want to place their water security in Turkey's hands or to bank on a technological solution that would be vulnerable to attack in so many countries.[17]

As with the other river basins in the region, the possibility of mutual gain from cooperation exists for all three nations dependent on the Tigris-Euphrates. Turkey, master of the vast majority of the rivers' flow, needs a water-sharing agreement with its riparian neighbors in order to secure financing from the World Bank and other lenders to complete the GAP, with investment costs estimated at some $29 billion. The benefits to Syria and Iraq of an agreement that builds water security are obvious. All three nations could gain from investments in irrigation efficiency and urban waterworks. Yet the Trilateral Commission on the Euphrates, which has met periodically, has discussed only rainfall data, flow levels, and other technical issues, leaving the tough political questions virtually untouched. Meanwhile, each nation pursues its own water strategy in an atmosphere of mistrust and insecurity.[18]

Although conditions for conflict are most ripe in the Middle East, international discord over scarce water exists elsewhere as well. In recent years, disagreements have heightened between Bangladesh and India over management of the Ganges River. In the early seventies, India completed the Farakka Barrage to divert water from the Ganges to the Hooghly River with the aim of improving navigation for the port of Calcutta. Newly independent Bangladesh was concerned that not enough Ganges water would cross into its territory during the dry season, causing its crop production to suffer.[19]

The two countries agreed in 1977 to a short-term solution for sharing the Ganges's dry-season flows, which gave Bangladesh 63 percent of the flow at Farakka and India 37 percent, with special provisions guaranteeing Bangladesh a minimum amount during extremely low-flow periods. This compact expired in 1982 and was replaced with an informal accord that did not include the guarantee clause for Bangladesh. A follow-up agreement expired in 1988. Since then, the two countries have been deadlocked, with neither agreeing to the other's long-term solution.[20]

Bangladesh's proposal includes building storage reservoirs on major tributaries of the Ganges that originate in Nepal. The dams would capture monsoon waters and help control flooding, as well as store water for irrigation during the dry season. India does not want to

involve Nepal, and has proposed instead the construction of a canal to channel water from the Brahmaputra
River to the Ganges, which officials in Dacca oppose. In
the absence of an agreement, India clearly has the upper
hand as the upstream nation, leaving Bangladesh at risk
of shortages.[21]

At the moment, international water law offers minimal assistance in resolving water conflicts. Upstream
countries have been reluctant to accept the notion of an
"international drainage basin" or "water system" that
should be cooperatively managed and equitably shared.
Indeed, some still hold the view that nations have "abso-
lute sovereignty" over water within their borders and
have little obligation to their neighbors.

An international code of conduct for shared watercourses has been steadily evolving, however, primarily
through the work of the private International Law Association, which in 1966 laid down the "Helsinki Rules on
the Uses of the Waters of International Rivers," and
through the United Nations International Law Commission, which in 1991 put forth its "Draft Articles on
the Law of Non-navigational Uses of International
Watercourses." Among the principles gaining favor are
the need to inform and consult with water-sharing
neighbors before taking actions that may affect them
(like Turkey's stoppage of the Euphrates's flow), to
avoid causing substantial harm to other water users, and
to allocate water from a shared river basin reasonably
and equitably. Each of these, however, is open to interpretation. For instance, the factors to be considered in
determining "reasonable and equitable" use are so
many and so broad as to offer little practical guidance.[22]

In the absence of a formal body of clear and enforceable law to which nations subscribe, resolution of water

disputes depends on the negotiation of treaties among neighboring countries. Notwithstanding the dismal picture in the Middle East, there have been a few successes on this front. Among the most notable is the Indus Waters Treaty, signed in 1960 by India and Pakistan, which had origins in a conflict that emerged when the subcontinent was partitioned in 1947. The international boundary delineating India and Pakistan cut right through the Indus River and the world's largest contiguous irrigation network, which encompassed 15 million hectares. The following year, as a way of claiming sovereign rights over the water within its territory, the Indian province of East Punjab stopped the flow of water into two large canals that fed Pakistan's irrigated land. The ensuing water dispute brought the two nations to the brink of war.[23]

Eight years of difficult negotiations preceded the signing of the Indus Treaty. The World Bank played a key mediating role by helping devise a water allocation strategy both nations deemed agreeable and by mobilizing international sources of funding for projects needed to implement it. Although the accord did not result in the basin being managed jointly and optimally by the two countries, it did partition water use equitably and establish a permanent commission to ensure its continued success. In force for more than three decades, the Indus agreement has helped promote regional development and political stability, and testifies to the mutual gain possible from the sharing of international waters.[24]

Progress has also been made in parts of sub-Saharan Africa. Since 1972, Mali, Mauritania, and Senegal have cooperated through a regional organization to manage the Senegal River. Although the large dams they have jointly constructed have worrisome environmental and social effects, the partnership approach these nations have taken to managing the river is a notable achievement. And under the auspices of the U.N. Environment Programme, the eight countries in the Zambezi River basin are working together to map out a river development plan that is both equitable and environmentally sound. Establishing such patterns of cooperation before scarcity and competition over water emerge can head off future conflicts.[25]

All told, the politics of water exhibit far more friction and strife than harmony and teamwork. No country can be economically or socially stable without an assured water supply. As more nations perceive water as an issue of national security, they may see that regional peace requires sharing it in an atmosphere of cooperation and trust.

7

A World
Heating Up

The prospect of planet-wide warming from the buildup
of carbon dioxide and other heat-trapping gases in the
atmosphere introduces a wild card into the global water
picture. Like a one-way filter, greenhouse gases allow
the sun's energy to pass through the atmosphere but
they trap the longer-wave radiation emitted back toward
space. As a result, the earth's temperature is expected to
rise, which in turn will alter the hydrological cycle—the
transfer of water between the sea, air, and land—in
some fundamental ways.

Climatologists project that the equivalent of a doubling of the concentration of heat-trapping gases over
preindustrial levels will raise the earth's temperature by
1.5–4.5 degrees Celsius (3–8 degrees Fahrenheit). The
warmer air will boost both evaporation and precipitation

globally by 7–15 percent. Rainfall patterns will shift, with some areas getting more moisture and others less. Hurricanes and monsoons are likely to intensify, and the sea level will rise from the warming of the oceans and the melting of polar ice caps.[1]

While global warming guarantees that the future will not be a simple extrapolation of the past, exactly how the climate will be different remains a puzzle. The major climatological models agree fairly well on global-scale changes, but they are not fine-tuned enough to predict accurately what will happen regionally and locally, or when. This uncertainty makes it difficult to plan new dams, reservoirs, and irrigation systems that are supposed to last a half-century or more. And, most disturbing, in cases where less rainfall is the outcome, areas already at or near water supply limits may move into an unbounded period of shortages.

To get a feel for the possible magnitude of effects in store, scientists have examined how the expected temperature rise—alone, and in combination with changes in precipitation—will affect the amount of water available in particular river basins. Their results are sobering, to say the least. An international group of scientists has concluded that a 1–2 degree Celsius warming along with a 10-percent decrease in precipitation—well within the realm of possibility in some areas—could reduce annual runoff by 40–70 percent. Such a drop would have staggering economic and environmental consequences on regions already short of water—forcing land out of irrigation, reducing hydroelectric power production, wiping out many species, and greatly constraining urban growth and the quality of life.[2]

Drought-stricken California offers a glimpse of what water shortages induced by a changing climate might bring. Although it is impossible to say whether the state's dry spell—now six years running—is related to global greenhouse warming, it serves as a useful analogue for thinking about the potential consequences of climate change.

In four of the five drought years that began in 1987, California's runoff—the volume of water in its rivers and streams—was less than half of normal. By May 1991, soil moisture had reached critically low levels, with the entire state south of San Francisco falling in the category of "extreme drought" according to a widely used soil-moisture index. California's many large reservoirs helped buffer the effects of sparse rainfall, but each year the drought persisted, reservoirs were drawn further down. By June 1991, the total volume of water in the state's reservoirs was at 55 percent of total capacity, and some were almost completely dry.[3]

As surface supplies diminished, water users began pumping more groundwater, particularly in the agriculturally important Central Valley. As a result, groundwater storage dropped as well, with water tables in seven Central Valley counties falling by 2–10 meters in less than five years.[4]

Each of these manifestations of water shortage has taken a severe toll on California's economy and environment. Following an in-depth study of the drought, Peter Gleick and Linda Nash of the Oakland-based Pacific Institute for Studies in Development, Environment, and Security found that low river flows have further decimated the declining striped bass population in the San Francisco Bay and delta, with the 1990 larval estimate the lowest ever recorded. Low flows have also destroyed the herring fishery in Tomales Bay, possibly irreversibly. Insufficient moisture has caused widespread tree mortality, with 30–80 percent of the trees in parts of the Sierra Nevada dead or dying in 1991. And a number of threatened and endangered plant and animal species are at greater risk of extinction because of drought-induced habitat and vegetation changes.[5]

Although the Californian economy has weathered six years of drought fairly well, some costs have been high, and the economic impacts will certainly magnify should the dry spell continue for several more years. Production of hydropower, which typically supplies about 20 percent of the state's electricity, dropped to about 12 percent of total electrical output, even as energy needs for pumping groundwater were increasing because of declining groundwater levels. To compensate, utilities burned more fossil fuels and purchased more power from other states, costing ratepayers about $3 billion since the drought began and putting more carbon dioxide and other pollutants into the air. Farm income suffered only slightly, with cash receipts dropping by about $400 million from an $18 billion total. But, as Gleick and Nash point out, some farm communities suffered disproportionately, and the entire sector will experience greater losses if the drought continues and groundwater reserves become too depleted to buffer cutbacks in surface irrigation supplies.[6]

While the California drought is a useful model of a future drier climate, it only illuminates part of the picture. Higher temperatures will characterize the greenhouse climate, whether local rainfall decreases or not. Warmer weather will increase evapotranspiration, and thus, other factors being equal, will boost water demands for crop irrigation, lawns and gardens, and other activities. More winter precipitation will fall as rain, and the snow that does fall will melt earlier in the year, reducing river flows during the summer when they are most needed. All told, the impacts of greenhouse warming could be far more severe than those of the ongoing California drought.

John Schaake, a hydrologist with the U.S. National Weather Service, analyzed potential changes in runoff for the Animas River at Durango, Colorado. He found that an increase of 2 degrees Celsius, with no change in precipitation, would affect total annual runoff very little. Seasonal runoff patterns, however, would change markedly as a result of reduced winter snowpack and faster melting of the snow that does fall. Schaake's model showed that average runoff for January through March would increase by 85 percent, while in the critical months of July through September it would drop by 40 percent. Such drastic shifts would increase the risk of winter flooding, reduce summer hydropower output, and lead to scarcities during the hot, dry summer, when water demands for crops and urban areas are greatest.[7]

It turns out that actual runoff measurements in northern California are beginning to show the kinds of changes that would be expected from a warming climate in that area—namely that spring runoff as a share of total yearly runoff is diminishing. From 1906 through 1980, runoff in the Sacramento River basin from April to July averaged 40 percent of the year's total. That average dropped to 33 percent between 1981 and 1990. Although it is not possible to link it to greenhouse warming, this shift is consistent with what would be expected and has likely worsened the consequences of the ongoing drought.[8]

Even where rainfall increases, it could bring problems along with benefits. Models suggest, for instance, that India might get more water, which could provide welcome relief to water-scarce parts of the country. But the additional rain could come in the form of more intense monsoons. If so, much of it would run off in damaging floods rather than augmenting soil moisture and stable supplies.[9]

Some agronomists harbor a less gloomy view of water resources in a greenhouse world than climatologists and hydrologists do. They see a world in which plants might need less water, since higher levels of carbon dioxide cause them to partially close their stomata, the leaf openings through which they exchange water vapor and other gases with the atmosphere. This enables them to use water more efficiently, yielding more plant matter per unit of water consumed.[10]

Studies have shown that doubling carbon dioxide levels can reduce transpiration—plants' consumption of water during photosynthesis—by a third to half. Across an entire watershed, this might substantially augment runoff from forests, grasslands, and croplands, possibly adding enough water to rivers and streams to counteract the effects of a drop in rainfall or increased evaporation. But this plant behavior has been demonstrated only in isolated, controlled experiments. Whether it would actually occur outdoors, where many other factors come into play, and whether it would be sustained rather than short-lived, remains unknown.[11]

All this uncertainty makes it difficult for farmers, water utilities, and engineers to know how to plan for the future and what kinds of investments to make. A farmer now depending only on rainfall to water crops is unlikely to buy an expensive irrigation system on the chance that rainfall may diminish in his or her region. Yet if it does, crop yields could plummet. Similarly, a water utility planning a city's supply for the next several decades faces the possibility of marked changes in sea-sonal runoff patterns, which could require different numbers and sizes of reservoirs, and different ways of operating them. If unprepared for the changes that occur, costly shortages could result.

As the earth warms over the coming decades, a mis-match is thus likely to exist between the need for water and the ability to supply it as reliably as we are accus-tomed to. Investment decisions of all kinds will remain hostage to uncertainties about just how the climate is changing in a given location.

For instance, based on what current models show, the Water Authority of Western Australia is anticipating a possible 20-percent drop in rainfall and a 45-percent reduction in river flows in the Australian southwest within 40 years. This would jeopardize a rich area of irrigated agriculture and the water supply for Perth. As a result, water officials have proposed that the federal gov-ernment spend $10 billion to construct an aqueduct that would transfer water from the Ord River on the north coast down toward Perth. But justifying such huge ex-penditures when regional climate changes are still so un-certain is difficult, to say the least.[12]

In agriculture, as in city water systems, adapting to a changing climate—once the direction and magnitude of change do become clear—will be costly. Some existing irrigation systems will no longer be needed, while other areas will require extra water to remain productive. For example, expanding irrigation to 5 percent of today's rainfed cropland to compensate for increased evapora-tion under hotter temperatures could cost $120–240 billion, or $12–24 billion annually for 10 years. This sum is on top of irrigation expenditures that would occur anyway to help meet the world's growing food

needs. For comparison, World Bank lending for irrigation has averaged roughly $1 billion per year since the
mid-eighties. Coming up with the needed funds would
be difficult. It seems likely that for some time irrigation
systems would be poorly matched to altered rainfall patterns and redistributed water supplies, making it more
difficult to grow enough food for the world's expanding
population.[13]

Most water officials in the United States, and presum-
ably elsewhere as well, appear to be taking a wait-and-
see approach to the prospect of global warming, choosing to do nothing until more evidence is in. In some
cases, however, a small adjustment made now as a
precautionary measure can avert much larger costs later
on. New York City, for instance, decided to increase the
height of a new drainage outlet in case sea level rises in
the future, a move that cost them nothing extra now, but
could prevent substantial commitments of capital down
the line.[14]

In other ways, too, we need not be totally hamstrung
by climate uncertainties. Improvements in irrigation
efficiency, urban conservation, and better regional coordination among water users—measures discussed in
Part II—can make water systems less vulnerable to
changes in climate that reduce water supplies. Californians, for example, are better positioned to deal with the
ongoing drought because many of them installed water-
efficient fixtures in their homes over the last 15 years. As
a result, many toilets and showers in that state take less
water than they otherwise would have, leaving a bit
more in shrinking reservoirs.[15]

In the end, efforts to stave off water scarcity and all its
consequences must include concerted action to slow the
pace of global warming. The climate treaty signed at the

June 1992 U.N. Conference on Environment and Development commits countries to control greenhouse gas emissions, but does not set targets and timetables for doing so. Water scarcity, food production, and climate change are now linked in complex ways. Although specific outcomes may be uncertain, it is clear that both water and food security will be more elusive for the next generation without rapid action to stabilize atmospheric greenhouse gases.

II

Living Within
Water's Limits

8

Thrifty Irrigation

With agriculture claiming two thirds of all the water removed from rivers, lakes, streams, and aquifers, making irrigation more efficient is a top priority in moving toward more sustainable water use. The savings possible in agriculture constitute a large and mostly unexploited new source of supply—the biggest pool in the "last oasis." Reducing irrigation needs by a tenth, for instance, would free up enough water to roughly double domestic water use worldwide.[1]

Amazing as it may seem, most of the world's farmers still irrigate the way their ancestors did 5,000 years ago—by flooding or channeling water by gravity across their cropland. Much water is lost as it is conveyed from reservoirs to farmlands, distributed among farmers, and applied to fields. Worldwide, irrigation efficiency is estimated to average less than 40 percent, which means the bulk of the water diverted for agriculture never benefits a crop. Although some of the "lost" water returns to streams or aquifers, where it can be tapped again, its quality is often degraded as it picks up salts, pesticides, and toxic elements from the land.[2]

Curbing agriculture's water use is also important since farming "consumes" a much larger share of the water devoted to it than industries or cities do. Because irrigation often takes place in dry regions, evaporation claims a substantial share. Together, evaporation and transpiration by the crops themselves use up about half the withdrawals from watercourses for agricultural use.

A wide variety of measures exist to boost agriculture's water productivity. They include new and improved irrigation technologies, better management practices by farmers and water officials, and changes in the institutions that govern the distribution and use of irrigation water. As discussed in Chapter 13, however, the vast water-saving potential of these measures will not be realized until the economic policies, laws, and regulations that shape decisions about water use begin to foster efficiency rather than discourage it.

Not surprisingly, some of the biggest technological successes in improving irrigation efficiency have occurred where water scarcity poses serious threats to farming. Two areas that stand out are the northwest corner of Texas, which is adapting to long-term depletion of the Ogallala aquifer, and Israel, which now boasts the most water-efficient agricultural economy in the world.

As in most places, a large share of land in the Texas High Plains is irrigated by channeling water along shallow trenches running parallel down a field. These so-called gravity or surface systems are typically the least expensive to install and are by far the most common method used in the world today. Unfortunately, most of them do not distribute water evenly. Farmers must often apply an excessive amount of water to ensure that enough reaches crops on the far side of the field. The wasted water percolates down through the soil, is lost to evaporation, or simply runs off the field.[3]

Farmers in Texas and elsewhere are turning to a technique called "surge" irrigation, which can greatly improve traditional gravity methods. Instead of releasing water in a continuous stream down the field channels, irrigation under the surge method alternates between two rows at specific time intervals. The initial wetting somewhat seals the soil, allowing the next application to advance more quickly down the furrow. This surging effect reduces percolation losses at the head of the field and distributes water more uniformly, especially if the furrows are somewhat shortened. Although the basic principle could be adapted for simple farming methods, surge units developed for the U.S. market include a valve and timer that automatically release water at pre-established intervals.

Farmers who adapt their old-fashioned furrow systems to the new surge technique have reduced their water use by 15–50 percent, while cutting their pumping costs at the same time. For those in the Texas Plains, where savings have averaged 25 percent, the initial investment (about $30 per hectare) is typically recouped within the first year.[4]

Especially in dry, windy areas, like the U.S. Plains, spraying water high into the air the way old-style sprinklers do results in very large evaporation losses and needless depletion of scarce groundwater. In recent years, many irrigators in northwest Texas have moved from high-pressure sprinklers, which typically register efficiencies of 60–70 percent, to low-pressure ones, which boost efficiency to around 80 percent. In a water district covering half the total irrigated land in the Texas panhandle, virtually all high-pressure systems have now been converted.[5]

A relatively new sprinkler design, known as low-energy precision application (LEPA), offers even greater savings. LEPA sprinklers deliver water closer to the crops by means of drop tubes extending vertically from the sprinkler arm. When used in conjunction with water-conserving land preparation methods, LEPA can achieve efficiencies as high as 95 percent. Since the system operates at low pressure, energy requirements often drop 20–50 percent. Adapting an existing sprinkler for LEPA costs Texas farmers in the range of $60–160 per hectare, and the water, energy, and yield gains typically make it a cost-effective investment. The payback for such a retrofit is two to four years, while converting to a LEPA sprinkler from an entirely different system would typically have a payback of three to seven years.[6]

Between the 1989 and 1990 growing seasons, Ken Carver and Robert George sold the conventional side-roll sprinkler that had been watering their alfalfa farm east of Lubbock, Texas, and bought a LEPA system. In addition, they buried gypsum blocks in the soil to monitor soil moisture, which allowed them to irrigate only when their crops really needed it. The detailed records they kept for those two years showed that water use dropped by 47 percent, electricity use fell by 32 percent, while crop yield—owing mainly to better-timed irrigations—rose by nearly a third. As a result, their overall water productivity—the amount of alfalfa produced with each cubic meter of water pumped—climbed 150 percent.[7]

Added up, efficiency improvements have helped stem the Ogallala aquifer's depletion. Since it peaked in 1974, water use in the Texas High Plains has fallen by 43 percent. Two thirds of this drop is due to cutbacks in irrigated area, but one third is attributed to conservation. On average, farmers used 16 percent less water per irrigated hectare during 1979–89 than they did in 1964–74. The average annual rate of Ogallala depletion has fallen from nearly 2 billion cubic meters per year during the late sixties to 241 million cubic meters in recent years, a drop of 88 percent.[8]

Necessity has mothered inventions in irrigated agriculture just as it has in so many other realms, and nowhere is this more evident than in Israel. Half desert, this Connecticut-sized nation has brought about what is widely perceived as an agricultural miracle over the last three decades. While it remains to be seen whether its success in making the desert bloom will prove sustainable, Israel has developed technologies, methods, and scientific capabilities in irrigation that could prove invaluable to much of the world as the era of water constraints unfolds.

Following its establishment as an independent nation in 1948, Israel faced the challenge of growing crops in a dry environment with extremely limited supplies of water that often contained high levels of salt. In response, Israeli researchers developed a new concept in agricultural water use known as drip irrigation—and they have been perfecting it ever since.

Under this method, water is delivered through a network of porous or perforated piping, installed on or below the soil surface, directly to the crops' roots. This keeps evaporation and seepage losses extremely low. Because water is applied frequently at low doses, optimal moisture conditions are maintained for the crop, which boosts yields, and salt is prevented from accumulating in the root zone. Modern Israeli farms often have highly automated drip systems, with computers and monitors sensing when and how much water to apply and determining the precise amount of nutrients to add. Israeli farmers now liken their irrigation practices to "feeding the plant with a teaspoon."[9]

After its commercial development in Israel in the sixties, drip and other "microirrigation" techniques began to spread. By the mid-seventies, a half-dozen countries—Australia, Israel, Mexico, New Zealand, South Africa, and the United States—were irrigating some land by drip methods, and drip area worldwide totaled about 56,600 hectares. Since then, its use has grown twenty-eight-fold, with nearly 1.6 million hectares watered by drip and micro-sprinklers in 1991. (See Table 8–1.)[10]

Drip systems often achieve efficiencies in the range of 95 percent. Farmers save energy, too, since water is applied not only in smaller amounts but at lower pressure. Because it is relatively expensive, however, with the initial outlay typically running $1,500-$3,000 per hectare, drip irrigation is mostly used on higher-valued fruit and vegetable crops, though more than 130,000 hectares of cotton, sugar, sweet corn, and other field crops are now watered by drip as well.[11]

The area under drip today represents only about half of 1 percent of world irrigated area, but some countries have moved rapidly toward these thrifty irrigation methods in recent years. The United States, for instance,

TABLE 8-1. *Use of Microirrigation, Leading Countries and World, 1991*[1]

Country	Area Under Microirrigation	Share of Total Irrigated Area Under Microirrigation[2]
	(hectares)	(percent)
United States	606,000	3.0
Spain	160,000	4.8
Australia	147,000	7.8
Israel[3]	104,302	48.7
South Africa	102,250	9.0
Egypt	68,450	2.6
Mexico	60,600	1.2
France	50,953	4.8
Thailand	41,150	1.0
Colombia	29,500	5.7
Cyprus	25,000	71.4
Portugal	23,565	3.7
Italy	21,700	0.7
Brazil	20,150	0.7
China	19,000	<0.1
India	17,000	<0.1
Jordan	12,000	21.1
Taiwan	10,005	2.4
Morocco	9,766	0.8
Chile	8,830	0.7
Other	39,397	—
World[4]	1,576,618	0.7

[1]Microirrigation includes primarily drip (surface and subsurface) methods and micro-sprinklers. [2]Irrigated areas are for 1989, the latest available. [3]Israel's drip and total irrigated area are down 18 and 15 percent, respectively, from 1986, reflecting water allocation cutbacks due to drought. [4]13,820 hectares (11,200 of them in the Soviet Union) were reported in 1981 by countries that did not report at all in 1991; world total does not include this area.

SOURCE: See endnote 10.

now waters 606,000 hectares by microirrigation methods, 3 percent of its irrigated area. At more than 70 percent, Cyprus boasts the biggest share of irrigated land under drip methods, which the tiny nation uses on its citrus, grapes, olives, nuts, and vegetables.[12]

Israel now has around half its total irrigated land under drip. Its efficiency improvements have reduced water use on each irrigated hectare by one third, even while crop yields have increased. A sobering addendum to Israel's remarkable achievement, however, is that since the rapid rise in water use efficiency during the seventies, further gains have been minimal. It remains to be seen whether technological advances can produce another quantum leap in water productivity on Israeli farms.[13]

A natural question is the extent to which Israel's modern approach to irrigation is transferrable to water-short areas with different cultures, farming practices, and experiences. One test has been carried out on the west bank of the Jordan River in the Jiftlik Valley. There, Bedouin families, with per capita incomes averaging less than $200 per year, had been irrigating some 400 hectares by traditional flood or furrow methods—with a water use efficiency estimated at less than 30 percent. With initial help from Israeli researchers, a combination of modern technologies was made available, including drip systems, improved seed varieties, and earthen water storage ponds. Financial assistance, in part from the Mennonite church, enabled farmers to invest in the new methods, and access to markets was assured so they could sell their produce.[14]

The results were striking. Between 1970 and 1985, as more and more Bedouin families saw the impressive results their neighbors were getting, adoption of the irrigation techniques spread rapidly. Irrigated area expanded tenfold—to 4,000 hectares—and, because of the drip method's efficiency, total water use did not increase. Per hectare yields rose three to five times. And because the boost in production required more labor for harvesting and marketing, employment increased as well. Although 40 percent of the land had initially been planted in grains (barley and wheat), the cropping pattern shifted to virtually all vegetable and plantation crops. Thus, the Jiftlik farmers now enjoy higher incomes from their sales of high-quality produce to Arab and European markets, while purchasing more of their basic foodstuffs.[15]

It may be too early to judge the overall effect of this technology transfer from an ecological, social, and cultural perspective. Yet the Bedouin farmers are now free of external assistance. Israeli irrigation specialist Uri Or, who was involved in the effort, notes that once the irrigation system is in place, it is easy to use and does not require a high degree of literacy or technical skill. Strictly in terms of water productivity, the gains made in the Jiftlik Valley over a 15-year period may be unparalleled anywhere in the world. Similar experiments have been carried out with Bedouins and other Arab communities in Gaza and the Negev, and they support the success in Jiftlik.[16]

New technologies that build efficiency into their designs—like surge, LEPA, and drip irrigation—can do much to make crop production less demanding of the world's water supply. Equally important, however, is raising the efficiency of the extensive surface canal systems that dominate the world's irrigated lands. In many developing countries, especially in Asia, improving the performance of canal systems is critical not just to conserving water, but also to raising crop yields closer to their potential and alleviating poverty.

Much land slated for irrigation, and often counted as irrigated, gets insufficient water or none at all because irrigation works are poorly maintained and operated. University of Sussex researcher Robert Chambers estimates, for example, that at least a quarter of the area declared irrigated in India yields far below its potential because of "tail-end deprivation"—lack of water toward the back of a canal system. "Tail-enders" suffer deprived livelihoods as a result, since equitable and reliable irrigation is key to raising incomes, generating employment, and, because it allows year-round crop production, promoting social stability.[17]

Ensuring more equal water distribution and discouraging excessive applications at the front of the network would allow more land to be irrigated with the same volume of water. Supplies from many large canal projects flow all night long, and, not surprisingly, much is wasted. Since farmers often do not know when the next delivery will come, they will usually take as much water as they can when they get the chance. As a result, those at the head of a canal system overwater their fields, while tail-enders wait in frustration for their share.

Many of these problems arise because irrigation officials rarely have any incentive to improve the performance of the systems they administer. Their operating budget, for example, may come from a state or national treasury and bear no relation to how well the system is functioning. Irrigation fees collected from the farmers may go back into a general treasury, rather than being used to operate and maintain that local system. And since farmers have little say in how their projects are managed and are not charged for water according to their use, they, too, have little incentive to use water wisely. In short, there is little accountability by those in control, and little control by those who are supposed to benefit.

The potential gains of correcting such deficiences are as large as the task is difficult. Chambers estimates, for instance, that management improvements in India could allow an additional 8 million hectares to receive irrigation water from existing canal projects. That would expand India's irrigated area by 19 percent, and perhaps double the yield from this newly irrigated land—without developing any new water sources.[18]

Besides being operated poorly, many Third World irrigation schemes give disappointing results because they were never completed, infrastructure is inadequate, or canals and other irrigation works badly need maintenance. In one farming area of Sri Lanka, for example, the main canal was built too low to supply the head of the project and too small to send water to the tail end reliably, preventing farmers from cultivating some 40 percent of the project area during the dry season. The World Bank estimated nearly a decade ago that failure to keep irrigation systems in good working order in Mexico had produced accumulated maintenance costs of about $3.5 billion—an average of $700 per irrigated hectare nationwide.[19]

Such pervasive shortcomings suggest that making existing irrigation projects function better will in most cases prove more cost-effective than building new water supply projects. But this is easier said than done. At a minimum, it involves bridging the gap between irrigation officials, who typically view their mission solely as supplying water, and farmers, who can only make optimum use of water if they have some control over it. Especially in government-run projects, some form of "water users association" is necessary for farmers to have some say in management decisions. Such an organization also provides a mechanism for collecting fees to cover operation and maintenance costs, and for involving farmers directly in maintenance activities. Many studies have shown that when farmers actively participate in projects and have some responsibility for their operation, canals and other infrastructure function better, a greater proportion of the project area gets irrigated, and crop yields rise.[20]

Field workers trained in techniques of social organization have fulfilled an important catalytic and coordinating role in the formation of water users associations, especially in many Asian countries. Spearheaded in large part by the International Irrigation Management Institute in Sri Lanka, this approach has been instrumental in effectively banding farmers together, improving communication between them and irrigation officials, and establishing workable patterns of operating and maintaining irrigation systems.[21]

In Nepal, for instance, "social organizers" began working in 1987 in the Sirsia-Dudhaura Irrigation System, one that had been plagued by many common management problems. Within a year or so, farmers had been mobilized to clean and maintain field channels, water schedules had been prepared, and many breaches in the system had been fixed. As a result, water efficiency increased, and the surplus water allowed more land to be planted in wheat the following season. And with more water reaching the tail end of the system, conflicts among the farmers themselves diminished.[22]

In Mexico, the federal government is in the process of turning over the operation, maintenance, and financing of 77 large irrigation districts to autonomous water users associations. By giving farmers more control over their water supplies and direct responsibility for the upkeep and management of irrigation projects, the government hopes to improve the productivity and financial status of these systems, which collectively cover some 3.2 million hectares, 62 percent of the nation's irrigated land. The plan is to turn over some 2 million hectares to water users associations by 1994. It remains to be seen whether farmers will be sufficiently organized and prepared to assume responsibility by this date. In any case, the Mexican plan is an important test worth watching.[23]

Water efficiency in agriculture can also be increased by jointly managing surface water and groundwater. Not all water used inefficiently is necessarily wasted. Seepage from unlined canals and other "lost" water often becomes part of a groundwater supply that can be tapped downstream. Frustrated "tail-enders" have sometimes dug or drilled their own wells to recapture excess water from the head of a canal network. More deliberately managing surface water and groundwater jointly can stretch available supplies and give more land a reliable and controllable source of water.

The Gangetic plain stretching across northern India, for instance, is underlain by an alluvial aquifer that is recharged by the extensive canal network that criss-crosses the region. Particularly in the western part of the plain, a proliferation of groundwater wells over the past couple of decades has greatly increased crop production from the available water supply. This type of "conjunctive use" takes careful management, however, to prevent both waterlogging and overpumping.[24]

Wherever farmers have control over their irrigation supply, whether in rich or poor countries, they can save a great deal of water by scheduling waterings to coincide more closely with their crops' needs. This requires periodically monitoring soil moisture and irrigating just before plants become stressed from lack of water. Farmers can do fairly well by extracting a soil sample and estimating moisture content by its consistency. A more convenient method is to use a moisture monitoring device, such as gypsum blocks, which, when buried in the root zone and connected to a meter, give a fairly precise reading of when a crop needs water. On test plots of alfalfa and corn, irrigation scheduling using gypsum blocks has led to a 14–27 percent reduction in water use compared with neighboring control plots.[25]

Californian farmers benefit from a state-run service called the California Irrigation Management Information System (CIMIS), which gathers data from 82 weather stations and provides readings on evapotranspiration, thus helping farmers calculate their crops' water needs. Farmers can get the information from local newspapers, radio programs, or soil conservation offices, or directly by hooking into the CIMIS computer. By comparing the evapotranspiration rate with their particular crops' water requirements, farmers can determine more precisely when and how much to irrigate. Greenhill Farms near Fresno has cut water use on its fruit trees by 35 percent just since 1988 by avoiding overwatering. Without doubt, the CIMIS program, which operates on an $800,000 annual budget, pays California taxpayers back handsomely in water savings.[26]

The most practical, economical, and appropriate way to raise irrigation efficiency will vary from place to place. But in most cases the means exist for farmers to cost-effectively cut their water use by 10–50 percent. Experience shows that an investment in irrigation efficiency is usually also an investment in the productivity of crops and soils. With better water management, yields often increase, erosion is reduced, and fertile cropland is less likely to become waterlogged, salted, or sapped of nutrients. Encouraging more widespread adoption of water-saving methods and making the institutional changes needed for better management would stretch scarce supplies, lessen ecological damage to overtaxed rivers and streams, and help farmers rich and poor get more value out of the resource.

9

Small-Scale Solutions

Once again in 1991, farmers in Mali sowed their seeds in spring with hopes that the rains would come to nourish them in May or June. Once again, as has happened more often than not during the last two decades, the rains let them down, holding back until early July and robbing them of a good harvest.[1]

Unreliable, uncontrollable, and insufficient rains often foil the best-laid plans of farmers in sub-Saharan Africa and other dry regions lacking irrigation. Although the 16 percent of world cropland that is irrigated constitutes the most productive agricultural sector, rainfed farming is the largest and, in many places, the neediest. In contrast to the U.S. Cornbelt, western Europe, and other areas where rains tend to be ample and reasonably reliable, the drylands of Africa, western India, north-central China, and southwestern Latin America present formidable challenges to crop production. Altogether, arid and semiarid lands cover about a third of the earth's land surface and are home to some 600 million people, including many of the world's poorest farmers. For them, conservation and more efficient use of scarce water is quite literally a matter of life and death.[2]

Hopes that large irrigation schemes will solve the water and food problems of these regions, particularly Africa's drylands, are fading fast. Only 4 percent of sub-Saharan Africa's cropland is currently irrigated, and most of it is in just four countries—Madagascar, Nigeria, South Africa, and Sudan. Few good dam sites remain, and the costs of large irrigation projects—in some cases $10,000–20,000 per hectare—make investments hard to justify. As World Bank water specialists Guy Le Moigne and Shawki Barghouti note, "Favorable conditions, such as good high-yielding aquifers, rivers with sustained year-round flows, and large tracts of irrigable lands are unfortunately not available to justify the type of massive investment that has gone into the Nile Basin, the Middle East, and Asia."[3]

But with farmers suffering crop failures in one out of every three years in many African countries, new approaches are desperately needed. Attention is turning now to the potential of smaller-scale projects—micro dams, shallow wells, low-cost pumps, moisture-conserving land techniques, and a wide variety of "rainwater harvesting" methods—to make food production, and life generally, more secure for dryland dwellers. Many of these efforts have proved more cost-effective and less disruptive to local communities than the large schemes that dominated development efforts during the past few decades. And because of their smaller size and use of local resources, they tend to be less damaging to the environment. Although small-scale projects cannot completely substitute for larger ones, they can play a much greater role than they do today.[4]

Improvements on many traditional methods show great potential for boosting crop production and stabilizing rural livelihoods in dry regions. One of the most widely heralded successes has taken place in the Burkina Faso province of Yatenga. There, increasing human numbers combined with diminishing rainfall levels during the seventies and early eighties confronted farmers with the stark choice of either coaxing more food out of their arid land or migrating from the area. With the aid of Oxfam, a U.K.-based development organization, Yatenga farmers turned to a traditional technique they had largely forgotten—building simple stone lines across the slopes of their fields to reduce erosion and help store moisture in the soil.[5]

They improved on the earlier technique by constructing the stone walls along contour lines, using a simple water-tube device to help them determine a series of level points. The stone lines, also known as bunds, cause rainwater to spread out over the land and slowly infiltrate the soil, rather than running off the field. Many farmers also built deep planting holes, called zai in the local Moore language, that collect and concentrate rainfall runoff around the plants. Together, the stone bunds and zai can raise yields 30–60 percent even in the first year, with additional increases possible in subsequent years as soil fertility builds. Equally important, the treatments help prevent total crop failure in the very dry years, greatly enhancing food security. By the end of 1989, 8,000 hectares in more than 400 Yatenga villages were benefiting from these techniques.[6]

A variation on this theme is to use a hearty grass as the barrier instead of stones. Vetiver grass, native to India and known there as khus, has proved a popular choice in some areas. When closely spaced along the contours of a sloping field, the grass forms a vegetative barrier that slows runoff, giving rainfall a chance to spread out and seep into the soil, much the same way as the stone bunds do. Over time, a terrace forms as sediment gets trapped behind the row of grass. As a result of the soil and water conserved this way, yields often increase by half.[7]

Where labor is abundant and can be mobilized, terracing is another effective way of capturing rainwater and raising productivity on sloping dry lands. By creating strips of flat land out of a hillside, terracing reduces erosion and, most important in dry regions, traps rain where it falls. Terracing is an age-old practice in many parts of the world. Pre-Hispanic terraced lands are estimated to have covered some 1 million hectares in Peru alone. Sadly, only about a fifth of this area is still cultivated today, the rest having been abandoned.[8]

One of the preeminent modern-day terracing success stories is in Kenya, a country with one of the world's highest population growth rates and arid to semiarid conditions across three fourths of its territory. With funds from the Swedish International Development Agency, the Kenyan government began working with local self-help groups in the Machakos District in southern Kenya in 1979. The existence of these traditional mwethya groups allowed the people of Machakos to participate fully in the project from the outset and provided a valuable organizational structure for carrying out the work. The vast majority of mwethya group members are women, who practice a method known as fanya-juu terracing, which basically involves digging a ditch and throwing the soil up-slope to form an earthen wall. Since the land between the lines of raised earth levels off over time, bench terraces naturally form. As with stone bunds and grass hedges, the terraces are constructed on the contour to maximize erosion control and rainwater conservation.[9]

The results have been impressive. Since the mid-eighties, the people of Machakos have built an average of 1,000 kilometers of terraces each year, and an estimated 70 percent of all cropland in the district is now terraced. The few studies that have examined production gains suggest that average corn yields on terraced lands increased by at least 50 percent. Though highly labor-intensive, with the burden falling mostly to women, terracing spread widely in large part because of its immediate benefits. And as Will Critchley emphasizes in *Looking After Our Land: Soil and Water Conservation in Dryland Africa*, "the existence of well developed self-help groups is one of the main reasons for the success of conservation activities in Machakos."[10]

As many as 3,000 years ago, agriculturalists in extremely dry places practiced techniques known as "runoff agriculture," in which rainwater is collected from a catchment area and then channeled to fields to provide enough moisture for crops to thrive in an otherwise hostile environment. These water harvesting methods allowed some ancient farming cultures to thrive where yearly rainfall averaged only 100 millimeters (4 inches). Over time, they became widely used throughout the Middle East, North Africa, China, India, northwest Mexico, and the American Southwest.[11]

In Israel's Negev desert, researchers have revived and improved on techniques practiced by the Nabateans, caravan traders who occupied the region 1,600–2,000

years ago. Just below the remains of the ancient Naba-
tean city of Avdat sits a 1-hectare farm watered by a
250-hectare catchment. Runoff is collected from a few
heavy rains during the wet season and diverted through
a network of channels to irrigate the fields. Encased by
small stone walls, the plots can store up to 250 millimeters of runoff, enough to keep crops moist throughout
the dry season.[12]

Researchers at the nearby Jacob Blaustein Institute
for Desert Research have developed a number of ways to
improve crop production with the limited water supply
available. Covering the soil between crop rows with
polyethylene, for instance, cuts down evaporation and,
by heating up the root zone, causes plants to use water
more efficiently. Yields of sorghum doubled with this
method, from about 1.5 tons per hectare to 3 tons. Research in the Negev has also shown that water harvesting
in extremely arid climates can support agroforestry—the
combined production of trees and crops. The trees draw
upon deeper soil moisture, while the annual crop ex-
ploits the upper layers. In this way, production of much
needed food, fodder, and firewood might be increased
in the driest lands.[13]

Virtually all the methods just described involve conserving or channeling rainfall to enhance the amount of
moisture available to crops. They are, in a sense, a form
of irrigation, since they supply more water to plants than
natural rainfall would alone. And they work: crop yields
in many cases increase by half or more. And yet, methods so dependent on rain do not give farmers the degree
of control over a source of water usually associated with
irrigation, or the reliability of supply necessary for high
and predictable yields. For this, an equally diverse set of
small-scale irrigation methods that store rainwater for

later application or that draw on local rivers, streams, or aquifers offer a way to boost crop production more consistently without the extensive environmental and social harm caused by large dam-and-diversion projects.

Small-scale irrigation systems are usually built and run by farmers themselves, so they frequently go uncounted in official government statistics. This overlooks what is often a vibrant and vitally important component of food production. In the Philippines, for example, the "informal" irrigation sector is estimated to account for about half the nation's irrigated area, and in Nepal, about three quarters. As Mark Svendsen and Ruth Meinzen-Dick of the International Food Policy Research Institute (IFPRI) point out, "While individually small, systems such as the zanjeras and other communal systems of the Philippines, the subaks in Bali, mountain systems of Nepal, tanks in Sri Lanka and Tamil Nadu, and indigenous smallholder systems in Africa collectively irrigate millions of hectares worldwide."[14]

In dry parts of southern India, capturing rainfall runoff and storing it in small or medium-sized reservoirs called tanks is an age-old practice. Tanks are still a central feature of the temple complex in most southern Indian villages. Even small tanks can provide life-saving irrigation to a field of crops withering from lack of rain. And supplementing rainfall with one well-timed irrigation can boost yields greatly. Studies in India's drylands have shown that one irrigation on otherwise rainfed lands can raise yields of sorghum and upland rice by 40–85 percent and can double yields of wheat. With most of India's cereals, pulses, and oilseed crops produced under rainfed conditions, encouraging the revival and improved use of tanks for supplemental irrigation in dryland areas could be extremely valuable.[15]

Although large-scale irrigation in Africa has proved costly, problematic, and prone to failure, traditional and small-scale efforts have often met with considerable success. Up until a few years ago, official irrigation statistics for sub-Saharan Africa excluded many of these methods, and when the U.N. Food and Agriculture Organization retroactively corrected its figures to include them, irrigated area (excluding South Africa) rose by 37 percent. Even now, however, some promising forms of indigenous irrigation do not get counted. More important, they do not get the investment credits, extension services, and other forms of support offered to public irrigation systems, which often have a poorer track record. As a result, small-scale irrigation's potential in Africa remains constrained and underdeveloped—and food production less secure.[16]

One such neglected method is garden irrigation—intensive cultivation of very small plots, usually by individual households. To ensure enough moisture, most garden cultivation takes place in wetlands, swampy areas, or other moist spots. In Zimbabwe, for instance, cultivators plant gardens in dambos, seasonally waterlogged lands at the head of a drainage system that collect runoff from nearby higher ground and channel it slowly toward the river into which the network drains. For irrigation during the dry season, families may draw water from shallow wells or nearby springs, or directly from the high water table under this land. Although intensive cultivation with heavy machinery would likely damage the soil and water regimes of these seasonal wetlands, recent research suggests that low-technology cultivation in dambos can be environmentally sustainable with proper land and water management practices.[17]

Though individually small in size, irrigated dambo

gardens together cover an estimated 20,000 hectares in
Zimbabwe, equal to 9 percent of the official irrigated
area of 220,000 hectares. Usually less than half a hectare, they are often too small to offer complete subsistence, but they employ labor when it is most available
and yield food when families most need it—during the
dry season. Indeed, during the 1986–87 drought, dambos were the only lands yielding corn in some communal
areas. And some exhibit remarkable diversity as farmers
exploit possibilities for multiple cropping: during one
season, one 2.5-hectare dambo plot in Zimbabwe
yielded a harvest of 23 different crop species and 26 tree
species, as well as bees, fish, reeds, and fodder grass.[18]

As with many indigenously developed small-scale irrigation schemes, lack of public assistance in the form of
credit, subsidized loans, and technical aid constrains the
potential role of dambo gardens in food production.
Capital costs to develop them range from $100–2,500
per hectare, depending on the technologies and type of
water pumps used. Even at the high end, these costs are
far less than for most large, formal irrigation projects in
Africa. But to individual families, the investment can be
a big hurdle. Whereas public irrigation systems are heavily subsidized, small dambo gardens get virtually no support. Loans and credit are hard to come by because
communal farmers often lack clear title to land, and because garden vegetables are not considered credit-wor-
thy. Women, frequently the cultivators of dambo lands,
face particularly severe obstacles to obtaining credit.[19]

Because dambo gardens have been neglected, more
research is needed on the potential environmental impacts of their wider use. But as IFPRI's Svendsen and
Meinzen-Dick point out, indigenously developed and
managed garden irrigation schemes are among "the

most promising success cases in African irrigation." Recognizing them as such, and revamping policies so as to better support and promote them, could do much to enhance food security among smallholders.[20]

In much of sub-Saharan Africa, the use of simple, low-cost wells and pumps to tap shallow groundwater or local rivers and streams offers great potential to increase food production and cash income for farm families. For instance, more than 100,000 hectares of Niger, as well as smaller areas in Chad, Mali, northern Nigeria, and several other Sahelian countries, are underlain by shallow groundwater that could be tapped in this way. Inexpensive, small-scale irrigation methods, developed and operated by farmers themselves and offering good control over a reliable water supply, have consistently proved more successful than the heavily subsidized large public projects often pushed by governments and development agencies.[21]

More than 100 million people in Africa could benefit from greater use of these methods. Unfortunately, the techniques have not spread widely in large part because farmers simply do not know about them or cannot gain access to them. Yet where the knowledge and technology have been made available, small-scale irrigation has taken off.

In northern Nigeria, farmers have adopted low-cost wells and pumps to tap shallow groundwater for irrigation in far greater numbers than expected. More than 8,600 wells had been established in three northern Nigerian states by the late eighties, each capable of irrigating up to 2 hectares of land. In contrast to the large government schemes, which have averaged $30,000 per hectare of public money, these small-scale efforts have cost farmers $1,000–2,000 per hectare, including the pumps. Yields have increased 25–40 percent in the wet season, and farmers have added a dry-season crop, giving them more cash and security.[22]

Similarly, around Gao and Timbuktu in Mali, where good local markets exist, farmers have augmented crop output with small and medium-sized pumps that can tap the nearby river or local groundwater. Thirty or more village schemes have sprung up around Gao alone, with impressive production results: yields of rice have averaged 4.8 tons per hectare. And in the Kanem area of Chad, a combination of shallow wells and portable pumps has allowed for an expansion of cultivated area and a 130-percent increase in farmers' cash income. Farmers like the portability of the pumps, which allows them to be moved as necessary and to be borrowed from neighbors in the case of a breakdown.[23]

As with the dambo gardens in Zimbabwe, most farmers invest in small-scale irrigation with little or no public assistance and, quite often, under conditions of restricted access to credit. For instance, farmers who wish to purchase a pump and well in one area of Bauchi state in Nigeria are required to belong to a qualifying cooperative (which only about a fifth do) and to make a down payment of 25 percent of the loan. The financing agency extends credit for only five months. As a result, many farmers cannot make the investment, and private irrigation expands far more slowly than it otherwise would. Moreover, where formal credit is unavailable or loan conditions too constrained, small-scale irrigation will be possible only for the relatively well-off, tending to widen the gap between rich and poor rather than helping alleviate poverty.[24]

Lack of land rights for women—who do most of the farm work in Africa—also inhibits irrigation's role in food production. Typically, women use their harvest not just to feed the family but to earn money for clothes, school fees, cooking materials, and other household needs. If irrigation is introduced and both the land rights and the crop are controlled by her husband, a good portion of a woman's income will go to him. Since he frequently will not spend his money on the family, many women will simply not do the work of irrigated farming. As Ellen Brown and Robert Nooter have observed, "Thus marginalized, women often withdraw almost entirely from irrigation."[25]

Smaller is not necessarily better—or benign. A thousand low-volume wells can overpump an aquifer as surely as a hundred powerful ones if developed without regard for the limits of the renewable groundwater supply. But small-scale projects using simple technologies have been a neglected stepchild in the realm of irrigation. Given proper attention and support, they can contribute far more, not just to expanding food production, but also to raising living standards among the poorest, making livelihoods more secure, and using local resources sustainably.

10

Wastewater
No More

In the hills of Israel's western Galilee lie the makings of an elegant solution to some of the world's most troubling water problems. There, small reservoirs receive sewage from Kfar Manda, an Arab village of 7,000. Managed by the neighboring Jewish community of Yodfat, the reservoirs detain, biologically treat, and store the wastewater for application on nearby cotton fields by drip irrigation. The Arab villagers get an inexpensive way to handle their sewage, which otherwise might flow untreated into their surroundings. And the Yodfat farmers get a reliable and less costly source of irrigation water containing enough nutrients to cut their chemical fertilizer needs markedly.[1]

This small project and many others like it throughout Israel do something most modern water engineering schemes fail to do: they treat wastewater as a resource to be used productively rather than as a nuisance to be gotten rid of. Water pure enough to drink serves many functions that do not require such high quality—including irrigating crops and lawns, manufacturing many kinds of industrial products, and flushing human waste into a sewer. Fresh water used once can be used again in the same home or factory (usually called recycling) or collected from one or more sites, treated, and redistributed to a new location (known as reuse). By better matching supplies of varying quality to different uses, more value can be derived from each liter taken from a river, lake, or aquifer—and the economic and environmental costs of developing new freshwater sources can be lessened. The challenge, in a nutshell, is to take the "waste" out of wastewater.

By far the greatest gains lie in redirecting water used in cities and towns for a second use on farms. Though typically viewed as pollutants, most wastewater constituents are nutrients that belong on the land, where they originated. Farmers worldwide spend heavily on chemical fertilizers to give their crops the nitrogen, phosphorus, and potassium that domestic wastewater contains in large amounts. According to one calculation, it takes the equivalent of 53 million barrels of oil—worth more than \$1 billion—to replace with fossil fuel-based fertilizers the amount of nutrients yearly discarded in U.S. sewage.[2]

By using municipal water supplies twice—once for domestic use and again for irrigation—would-be pollutants become valuable fertilizers, rivers and lakes are protected from contamination, the irrigated land boosts crop production, and the reclaimed water becomes a reliable, local supply. Unfortunately, conventional sanitary engineers tend to emphasize the linear approach to managing water and sewage—use, collect, treat thoroughly, and then dispose of—while the benefits of closing the cycle—use, collect, treat partially, and then use again—go unrealized.

The idea of applying wastewater to cropland is not new. "Sewage farms" were operating in Edinburgh, Scotland, as early as 1650 and soon thereafter sprung up outside London, Manchester, and other English cities as well. Werribbee Farm in Melbourne, Australia, began functioning in 1897 and today irrigates about 10,000 hectares with reclaimed wastewater from its treatment ponds, the largest system of its kind in the world.[3]

Earlier this century, however, using treated wastewater to irrigate cropland fell out of favor as expanding cities encroached on the sewage farms and concerns arose about the transmission of disease from consuming vegetables irrigated with raw wastewater. With the development of modern biological and chemical treatment methods, which require much less land, sewage farming gradually became viewed as unsanitary, antiquated, and not worthy of further study—and was almost completely rejected in urban parts of industrial countries. Slowly over recent decades, however, it has come back, not only as an economical method of pollution control but, increasingly, in response to water scarcity.[4]

At least 500,000 hectares of cropland in some 15 countries are now being irrigated with municipal wastewater. Although this amounts to two tenths of 1 percent of the world's irrigated area, in dry regions wastewater can make up an important share of agriculture's water supply.[5]

Israel has the most ambitious wastewater reuse effort under way in the world today. Already, some 70 percent of the nation's sewage gets treated and reused to irrigate 19,000 hectares of agricultural land. With no new sources of fresh water to tap, Israel plans to expand the use of reclaimed wastewater greatly by the end of the decade. Virtually all of it will go to agriculture, which is projected to lose as much as 38 percent of its allocation of fresh water to spreading urban areas. If the nation meets its targets, reclaimed wastewater will supply more than 16 percent of Israel's total water needs by the end of the nineties.[6]

About half of Israel's reclaimed water comes from the greater Tel Aviv metropolitan area, where it undergoes treatment, is recharged to an underlying groundwater basin, is detained there for further treatment, and then is pumped back up and piped to farms in the western Negev desert. Although this scheme is the most lauded of the nation's reuse projects, it is not the one that most excites Shaul Streit, the director of Israel's Sewerage Project, who views it as "redundant" and unnecessarily costly. Streit prefers the simple elegance of the "agro-sanitary" approach used in parts of the western Galilee—partial treatment in wastewater lagoons and reservoirs followed by irrigation—to solve simultaneously problems of pollution, water scarcity, and crop production in dry lands.[7]

The key to this low-tech strategy is a series of ponds and reservoirs that biologically treat sewage and remove its harmful constituents, making it safe for watering crops that will not be eaten raw. Research has shown that, with a sufficiently long detention time, the anaerobic and aerobic treatment given by these lagoons can eliminate dangers from disease-causing organisms, such as bacteria, viruses, and parasitic worms. Organic matter is reduced so as not to overload the land, but a share

is retained to add useful nutrients and other elements to
the soil. The typical concentrations in wastewater would
give most crops the nitrogen they need, and much of the
phosphorus and potassium as well. Studies in California, Israel, and Portugal have shown that many crops
irrigated with wastewater do very well with no additional
chemical or organic fertilizer.[8]

One obvious drawback to this approach is the need
for a sizable area of flat land on which to construct the
treatment ponds near the city or town. Adequately treating the wastewater generated by 100,000 people in a
fairly warm climate would require about a 30-hectare
lagoon. Where land is expensive, purchasing the needed
area can increase the cost of these systems substantially.
But rarely will they be more expensive than treatment by
a conventional sewage plant. Not counting land costs,
pond systems typically run about $4 per person a year,
compared with a minimum of $25 per capita annually
for conventional systems. High land costs might bring
yearly pond treatment costs to $10–16 a head, still a
bargain. And when the added benefits from irrigation
are counted, the economics of the "agro-sanitary" approach become even more favorable.[9]

Another problem to overcome is that wastewater is
generated year-round, while the irrigation season may
last only a few months. Some means of storing the
wastewater is needed to get the maximum irrigation
benefits from the system and to prevent partially treated
sewage from being released to the environment. Assuming an average of 120 liters of wastewater is released to
the sewer system per person a day, a city of 500,000
people would be able to supply enough treated effluent
to irrigate some 2,700 hectares—but only if the
wastewater can be stored.[10]

Israel has taken to building deep reservoirs coupled with the lagoons to provide this needed storage, and now has more than 120 seasonal wastewater treatment reservoirs in operation. The nation also has the advantage of irrigating much of its cropland by the highly efficient drip method (see Chapter 8), which means it can irrigate a greater area—and thus get more benefit—from the wastewater it generates. And since drip systems deliver water directly to the crops' roots, health risks from spraying or channeling reclaimed wastewater do not arise.[11]

When designed and operated properly, waste stabilization ponds offer a low-cost way to keep sewage out of rivers and streams, safeguard human health from disease-causing organisms, and produce a nutrient-rich source of irrigation water. Studies have shown them capable of treating wastewater up to the World Health Organization's standards for irrigation of crops not eaten raw. These standards include limits on fecal coliforms and parasitic worms, two of the most important disease-causing agents. Care must always be taken to prevent damaging levels of heavy metals from getting into wastewater destined for irrigation. Cadmium, copper, nickel, zinc, and other heavy metals can accumulate in crops and soils, or percolate to groundwater and contaminate a drinking supply. A key to safe reuse is thus preventing untreated industrial effluent, which often contains heavy metals, from mixing with domestic wastewater.[12]

Unfortunately, the wastewater reuse practices of many developing countries are far from safe and sanitary. Most of the wastewater flowing from the urban centers gets no treatment, and in water-short areas, it is often applied raw to edible crops. Raw sewage from

Santiago, Chile, for instance, makes up almost the entire flow of the Rio Mapocho during the dry season. This water irrigates about 16,000 hectares of vegetable and salad crops headed for city markets, a practice linked to typhoid fever outbreaks in Santiago in the mid-eighties.[13]

Similarly, irrigation districts outside Mexico City apply raw sewage channeled from this huge metropolis to their fields. One district in the southwestern state of Hidalgo receives about 3.5 million cubic meters of raw sewage each day. Although farmers are prohibited from using this water to irrigate crops that are consumed raw, the prohibition is apparently not always respected or enforced. Some vegetables have been found to be highly contaminated with fecal coliforms, posing a direct threat to human health.[14]

By not making wastewater reuse a part of water planning and management, developing countries put their urban and rural populations at risk. As World Bank wastewater specialists Carl Bartone and Saul Arlosoroff note: "In dry climates, if reuse is not planned and reuse policies not defined, reuse will almost surely take place anyway out of economic necessity—but without adequate sanitary controls. Examples abound of local farmers breaking into sewer interceptors both within and on the outskirts of urban areas to steal the effluents for watering their crops. These are often vegetable crops destined for local markets that will be consumed raw. In addition, indirect reuse *is* occurring everywhere. Highly polluted rivers serve as major water sources for large-scale irrigation projects."[15]

Most of the wastewater reclamation going on in industrial countries is of a higher-tech variety designed to meet more stringent quality standards. Often patterned after those set in California, these standards essentially require that reclaimed wastewater be virtually free of disease-causing organisms before it is used to irrigate agricultural food crops, parks, or playgrounds. Standards are a bit less strict for irrigating pasture, golf courses, and greenbelts, and they are least strict for irrigation of fiber, forage, orchard, and vineyard crops, where direct human exposure to the wastewater is minimal.[16]

Typically, engineers achieve the required treatment through a sequence of physical, biological, and chemical processes that collectively can bring wastewater up to a very high quality. Depending on the size and type of operation, the advanced treatment needed to meet the strictest standards for reuse costs 15–42¢ per cubic meter ($180–520 per acre-foot), including conventional primary and secondary treatment. Although expensive, this costs less than developing new water sources in much of the western United States, for instance, and is much cheaper than seawater desalination.[17]

As in Israel, advanced wastewater treatment and reuse is expanding in water-short parts of the United States, though at a slower pace. Los Angeles plans to be reusing 40 percent of its municipal wastewater within 20 years. Its "Water Factory 21" has been reclaiming sewage for 15 years in order to recharge underlying aquifers and thereby combat the intrusion of seawater into coastal groundwater reserves. Long-term plans in Tucson, Arizona, call for reclaimed wastewater to meet 19 percent of total water needs. The city's present level of reuse (7.4 million cubic meters per year) is expected to roughly quadruple by the end of this decade, with a good portion of the reclaimed water being stored for later use in underground aquifers, which have suffered extensive depletion.[18]

Phoenix, another water-strapped Arizona city, has worked out an interesting agreement with irrigators and the U.S. Bureau of Reclamation involving an exchange of treated wastewater for fresh water. In effect, the city will get roughly 2,000 cubic meters of fresh water for every 3,000 cubic meters of treated wastewater it ships to a nearby irrigation district. The agreement allows for the transfer of up to 37 million cubic meters of reclaimed water each year, potentially giving Phoenix more than 24 million cubic meters of drinking-quality water. The facilities needed to carry out the transaction are expected to be in place by the end of 1993.[19]

St. Petersburg, Florida, is apparently the only major U.S. city to have closed its cycle completely by reusing all its wastewater and discharging none to surrounding lakes and streams. The city has two water distribution systems—one that delivers fresh water for drinking and most household uses, and another that distributes treated wastewater for irrigating parks, road medians, and residential lawns, and for serving other functions that do not require drinking-quality water. For residents hooked up to the dual system, the reclaimed water costs only about 30 percent as much as the drinkable supply, and, because of the nutrients it contains, cuts down on their lawn-fertilizer costs as well.[20]

Projects in a few cities in the world are pioneering the reclamation of municipal wastewater for drinking. Windhoek, Namibia, was the first city to add reclaimed water to its public supply, and has been doing so for more than 15 years. In the United States, Denver, Colorado, has completed a pilot project demonstrating the technical feasibility of reclaiming wastewater to drinking quality. And El Paso, Texas, is injecting highly treated wastewater through wells into an aquifer, where it travels 3 kilometers downstream for two to four years before it is recovered by the city's water supply wells.[21]

Drinking water produced through both the El Paso and the Denver projects is expensive—in each case, more than 70¢ per cubic meter ($864 per acre-foot). This makes reclamation cost-competitive, or nearly so, with the development of new freshwater sources in parts of the West, but a host of conservation measures and efficiency improvements remain far more economical. (See Chapter 12.) In general, turning wastewater into drinking water makes less sense than using it for irrigation, which does not require such costly treatment.[22]

In sum, the major barriers to reusing wastewater are psychological, not technical. The greatest gains from it would come from larger scale reuse of municipal wastewater in agriculture. A major push by development agencies, governments at all levels, and private engineers to combine low-cost sewage treatment with irrigation could go a long way toward solving the vexing triad of pollution, scarcity, and health problems now plaguing so much of the world.

11

Industrial Recycling

Making the myriad products we use in everyday life—from clothes and computers to paper, plastics, and televisions—requires copious amounts of water. Producing 1 kilogram of paper can take as much as 700 kilograms of water. And making a ton of steel can take 280 tons of water.[1]

Collectively, industries account for nearly a quarter of the world's water use. In most industrial countries, they are the biggest user—frequently accounting for 50–80 percent of total demand, compared with 10–30 percent in much of the Third World, where industrial use typically trails that of agriculture by a wide margin. As developing countries industrialize, however, their water demands for electric power generation, manufacturing, mining, and materials processing are rising rapidly.[2]

In contrast to the water used in agriculture, only a small fraction of industrial water is actually consumed. Most of it is used for cooling, processing, and other activities that may heat or pollute water but do not use it up. This creates the possibility of recycling supplies within a factory or plant, thereby getting more output from each cubic meter delivered or allocated to that operation. U.S. steelmakers, for instance, have reduced their water intake to 14 tons per ton of steel, securing the remainder from recycling.[3]

So far, the main impetus for industrial water recycling has come from pollution control laws. Most of the world's wealthier countries now require industries to meet specific water quality standards before releasing wastewater to the environment. As it turns out, the most effective and economical way to comply with pollution requirements is often to recycle and reuse water a number of times. Pollution control laws have therefore not only helped clean up rivers, lakes, and streams, they have promoted conservation and more efficient water use.

Japan, the United States, and the former West Germany are among the countries that have achieved striking gains in industrial water productivity. After rapid industrialization following World War II, total water use by Japanese industries peaked in 1973 and then dropped 24 percent by 1989. Three industries—chemicals, iron and steel, and pulp and paper manufacturing—account for 60 percent of Japan's industrial water use, and each has boosted its water recycling rate markedly since the early seventies. Industrial output, meanwhile, has been climbing steadily. As a result, in 1989, Japan got $77 worth of output from each cubic meter of water supplied to industries, compared with

$21 of output (in real terms) per cubic meter in 1965. In just over two decades, the nation more than tripled its industrial water productivity. (See Figure 11–1.)[4]

A similar picture emerges in the United States. U.S. industry's total water use has fallen 36 percent since 1950, while industrial output has risen 3.7-fold (in real terms). The bulk of the water used in U.S. manufacturing (excluding thermoelectric power production) occurs in four industries—paper, chemicals, petroleum, and primary metals. In response to increasingly strict pollution control regulations, each has steadily increased its water recycling rate. Whereas U.S. manufacturing operations were using each liter of water supplied to them an average of 1.8 times in 1954, the recycling rate is predicted to rise to 17 by the end of the nineties. Unfortunately, this remains somewhat speculative; no census

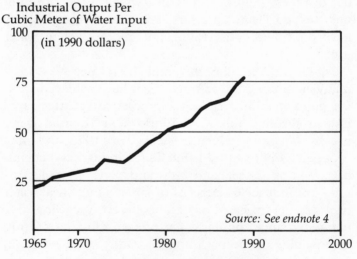

FIGURE 11-1. *Industrial Water Productivity in Japan, 1965–89*

has been taken of water use in manufacturing since the early eighties, and so trends over the last decade are not documented.[5]

In the former West Germany, total industrial water use today is at the same level it was in 1975, while industrial output has risen 44 percent. State-of-the-art paper manufacturing plants there now use only 7 kilograms of water to produce a kilogram of paper, 1 percent as much as older factories elsewhere. In Sweden, too, strict pollution control requirements have led to widespread adoption of recycling in the pulp and paper industry, the nation's biggest water user. Between the early sixties and late seventies, the industry cut its water use in half while doubling production—a fourfold increase in water productivity.[6]

In deciding how much to recycle, a manufacturing plant weighs the costs of getting water and treating it prior to disposal with those of adding equipment to treat and reuse wastewaters within the plant. In most industries, recycling partially offsets its costs by recovering valuable materials, such as nickel and chrome from plating operations, or fiber from papermaking. As water supply and wastewater treatment costs rise, recycling becomes more cost-effective. And in water-short regions, industries increasingly turn to recycling to guard against possible cutbacks in supply. The response of the most innovative companies to these various incentives shows clearly that further large cuts in industry's water needs are possible.

Spalding Sports Worldwide, a sporting goods manufacturer, has a Massachusetts plant that draws upon the Quabbin Reservoir, the main source of supply for the greater Boston area. When the region's brush with drought in 1988 and 1989 further squeezed an already tight water supply outlook, the company decided to take a hard look at its entire production process with an eye toward expanding the conservation efforts it had begun earlier in the decade. Through a number of measures—especially recycling the cooling water for its machinery—Spalding cut its water use from 1.5 million cubic meters in 1989 to 64,000 cubic meters in 1992—a 96-percent drop in just three years.[7]

For Epton Industries, a Canadian manufacturer of rubber and plastics, the main incentive for water recycling was a 1990 bylaw adopted by the Regional Municipality of Waterloo that made it illegal to discharge to the sewer system cooling water that was used only once. Like many manufacturers, Epton uses water to cool its products during processing. This often requires a very high volume, but the water never touches the product itself. In response to the new law, Epton began recirculating and reusing its cooling water by simply chilling it back down to a low temperature. This, along with other conservation measures, reduced its water use by 60 percent within a year.[8]

As part of a city-wide conservation effort during the mid-eighties (see Chapter 12), industries in San Jose, California, made impressive progress in water conservation. A detailed look at 15 companies in the area—including several computer makers, a food processer, and a metal finisher—showed that by adopting a diverse set of conservation measures these firms collectively reduced their annual water use by 5.7 million cubic meters—enough to supply about 9,200 San Jose households. Water savings ranged from 27 to 90 percent, and in most cases, the payback period on the conservation investments was less than 12 months. (See Table 11–1.)[9]

TABLE 11-1. *San Jose, California: Industrial Water Conservation and Cost-Effectiveness, Selected Companies*

Company	Water Use Before Conservation	Water Use After Conservation	Water Savings	Payback Period on Investment
	(thousand cubic meters per year)		(percent)	(months)
IBM[1]	420	42	90	3.6
California Paperboard Corp.	2,473	689	72	2.4
Gangi Bros. Food Processing	568	212	63	10.8
Hewlett-Packard[1]	87	42	52	3.6
Advanced Micro Devices	2,098	1,318	37	7.2[2]
Tandem Computers	125	87	30	12.0
Dyna-Craft Metal Finishing	193	140	27	2.4

[1]Water use rates apply only to one or more processes involving conservation measures. [2]Payback based only on that portion of water savings with which costs could be associated.

SOURCE: See endnote 9.

Indeed, one positive outcome of California's six-year drought is that this state, which has an economy larger than all but seven countries, may now be the world's leader in industrial water recycling. Manufacturers of many kinds have boosted their water efficiency dramatically in a matter of years. Besides the typical inducements of strict federal and state water quality regulations, Californian industries have faced the possibility of large cutbacks in water supply because of the ongoing drought. As a result, many are investing in water conservation well beyond what is financially justified at the present time as an insurance policy against future rationing, which could threaten production.[10]

A 1990–91 survey of 640 manufacturing plants in 12 California counties showed that these industries saved some 94 million cubic meters of water in 1989 compared with 1985, a 19-percent reduction and equivalent to the annual use of 150,000 households. These savings came on top of impressive conservation gains made during the previous 15 years in response to increasingly strict environmental standards. All told, the three largest water-using industry groups in these counties have cut their water demands by nearly two thirds during the last two decades.[11]

Measures to recycle cooling and process water form the backbone of the water-saving investments these industries have made. These span a wide range of engineering designs and costs, and constitute both some of the least-cost, simplest measures and some of the most complex and expensive. Other improvements included changing water nozzles to reduce flow rates, switching from continuous to intermittent flows in some manufacturing processes, sequential reuse of process water, and monitoring for leaks.[12]

Six industries outperformed the others—aircraft, computers, electronic components, fruit and vegetable processing, motor vehicles, and paint. Collectively, these firms used only about half as much water for each unit of production in 1989 as they did in 1985. Makers of office and computer equipment, for example, increased their output by 56 percent over this five-year period while cutting their total water intake by 21 percent—nearly doubling production per cubic meter of water used. Moreover, if all Californian plants came up to the level of the most efficient ones of their type, total water needs in all the industry groups surveyed would drop another 19 percent.[13]

Unfortunately, few developing countries are yet giving industries the incentives they need to adopt more efficient water practices. Most neither charge appropriately for water and wastewater services nor enforce pollution control regulations adequately. In Egypt, for example, 117 factories—including dozens of textile, chemical, and metal companies—drain their wastewater directly into the Nile River, the country's sole source of drinking water. Some 343 factories along the banks of South Korea's Naktong River illegally discharged toxic wastes in 1990. And the Rimac River in Peru, which supplies 60 percent of Lima's drinking water, is contaminated with arsenic, chromium, cyanide, and other toxic elements in concentrations up to twice those considered safe by the World Health Organization.[14]

Besides allowing water supplies to become dangerously contaminated, the failure to control industrial pollution and thereby encourage recycling causes industry's water demands to climb rapidly. Even while industrial water use is leveling off or falling in Japan, the United States, and western Germany, it is projected to increase markedly in most of the developing world. This will further strain the capacity of water supply systems already stretched to meet the demands of rapidly growing cities.

Despite the temptation to promote economic growth through heavy subsidies for industrial water and lax enforcement of pollution standards, a number of cities in the developing world are actively promoting industrial conservation and recycling. In Singapore, for instance, industries pay more per cubic meter than households do, a departure from the more usual practice of keeping industrial rates low as an enticement to manufacturers. Singapore also imposes a 15-percent water conservation tax on industries using more than a specified amount. New factories that will require more than 500 cubic meters of water per month must get city approval before they start operating, and officials work with them from the planning stages to help them incorporate conservation, recycling, and, where possible, use of lower-quality water in their operations.[15]

In the town of Goa, India, about 380 kilometers south of Bombay, a fertilizer plant owned by Zuari Agro-Chemical Limited cut its water use by half over six years in response to high water prices and government pressure to reduce effluent discharges to the sea. The Goa plant now produces a ton of nutrients using only 40 percent as much water as a fertilizer factory at Kanpur in Uttar Pradesh. Similarly, in São Paulo, Brazil, high effluent charges encouraged a dairy factory, a pharmaceutical company, and a food processing plant to reduce their water use per unit of output by 62 percent, 49 percent, and 42 percent, respectively.[16]

Given the proper incentives, industries of many types have shown they can cut their water needs 40–90 percent with available technologies and practices, while at the same time protecting water from pollution. Industrial conservation offers many cities facing shortages a large untapped new supply. By enforcing pollution control requirements and with aid from donor countries and development agencies, countries that are now rapidly industrializing can "leapfrog" to these resource-saving techniques.

Ensuring that new factories incorporate conservation and recycling from the outset would help delay costly investments in urban water supplies, reduce overpumping of aquifers, lessen competition for water, and help prevent pollution from reaching levels hazardous to people and wildlife. Closing the industrial water and wastewater cycle is not only technically possible, it increasingly makes good economic and environmental sense.

12

Conserving
in Cities

Mexico City's historic plaza offers up a strange sight. The imposing Metropolitan Cathedral, built soon after the sixteenth-century Spanish conquest, droops rather dramatically on its right-hand side, less so on the left. Inside, an array of tension wires and green metallic beams support the weakening edifice. The capital's revered cathedral is sinking, and the reason has little to do with the engineering skills of the early Spaniards. Large parcels of land are subsiding as the city depletes its groundwater, and the resulting structural damage is but the most visible of the consequences.[1]

Mexico City, while an extreme case, is just one of numerous cities around the world that have overstepped water's limits. Homes, apartments, small businesses, and other municipal enterprises account for less than one tenth of the world's total water use, but their demands are concentrated in relatively small geographic areas and, in many cases, are escalating rapidly. As cities expand, they strain the capacity of local water bodies and force engineers to reach out to ever more distant sources.

In addition, the reservoirs, canals, pumping stations, pipes, sewers, and treatment plants that constitute a modern water and wastewater system require large sums of money to build and maintain. Collecting and treating water and wastewater also takes a great deal of energy and chemicals, adding to environmental pollution and the overall costs of a community's water system. Under such constraints, many cities are having difficulty meeting the water needs of their residents, and large numbers of low-income residents in developing countries get no service at all.

Conservation, once viewed as just an emergency response to drought, has been transformed in recent years into a sophisticated package of measures that offers one of the most cost-effective and environmentally sound ways of balancing urban water budgets. Just as energy planners have discovered that it is often cheaper to save energy—for instance, by investing in home insulation and compact fluorescent lights—than to build more power plants, so water planners are realizing that an assortment of water efficiency measures can yield permanent water savings and thereby delay or avert the need for expensive new dams and reservoirs, groundwater wells, and treatment plants. Slowly the idea is spreading that managing demand rather than continuously striving to meet it is a surer path to water security—while saving money and protecting the environment at the same time. (See Table 12–1.)[2]

TABLE 12-1. *Urban Conservation Initiatives, Selected Cities*

City/Region	Activities/Accomplishments
Jerusalem, Israel	Installation of water-saving devices, leak detection and repair, and more-efficient irrigation of parks contributed to a 14-percent drop in per capita use from 1989 to 1991.
Mexico City	Replaced 350,000 toilets with 6-liter models, saving enough water to meet needs of 250,000 residents; goal of cutting per capita use by one sixth by 1996 through pricing, education, retrofitting, and efficiency standards.
Southern California	Metropolitan Water District pays member agencies $125 for each 1,000 cubic meters they save. Estimated savings as of June 1992 total nearly 33 million cubic meters/year. Conservation efforts have cut annual demand by 541 million cubic meters, enough to supply about 885,000 households.
Beijing, China	New pricing system links charges to amount of water used; regulations from November 1992 set quotas on consumption and authorize fines for exceeding them.

City/Region	Activities/Accomplishments
Singapore	With water use rising more than twice as fast as population, the island nation cut unaccounted-for water to 10 percent through leak repairs, and promoted conservation with higher water prices and public education.
Greater Boston, Massachusetts	Comprehensive retrofit, water audit, leak detection, and education program reduced total annual demand by 16 percent, bringing it to the level of late sixties.
Municipality of Waterloo, Canada	Delayed expansion of regional water supply through higher water rates, distribution of water conservation kits, and public education. Per capita water use has fallen 10 percent during last three years.
Bogor, Indonesia	Faced with costly water supply project, utility hiked water rates to encourage conservation. Within a year, average monthly residential water use dropped nearly 30 percent.
Melbourne, Australia	Since 1982–83 drought, when water use dropped 30 percent, a conservation strategy has kept water use from climbing above the level of 1980, allowing construction of new water works to be postponed and saving $50 million.

SOURCE: See endnote 2.

Many urban areas simply have no feasible way of balancing supply and demand without conservation and more-efficient water use. Mexico City epitomizes such a case. This sprawling metropolis of some 18 million people currently relies on groundwater for more than 80 percent of its supplies. Pumping exceeds natural recharge by 50–80 percent, which has led to falling groundwater levels, compaction of the aquifer, subsidence of the land, and damage to surface structures, including the cathedral.[3]

The city sits in a geographic bowl higher than the surrounding countryside, so its only possible surface water sources lie far below. Having outstripped the limits of local groundwater, the city now meets 17 percent of its demand by bringing water from the Cutzamala River system 127 kilometers away and lifting it 1,200 meters in elevation—incurring a huge energy cost. With the metropolitan area expanding by more than a half-million people each year, officials are racing against time to achieve some degree of water stability.[4]

Faced with such an intractable problem, the Mexican government and city officials are orchestrating an aggressive water conservation initiative. In 1989, the federal government took a bold step and adopted a strict set of nationwide efficiency standards for household plumbing fixtures and appliances. They require toilets—the biggest water guzzler in the home—to use no more than 6 liters (1.6 gallons) per flush, and they set maximum limits for showers, faucets, dishwashers, and washing machines as well.[5]

Mexico City has launched an ambitious program to replace conventional toilets (using about 16 liters) with the 6-liter models in public places, commercial buildings, and private residences. By late 1991, more than

350,000 toilets had already been upgraded, which will save nearly 28 million cubic meters of water per year—enough to meet the household needs of more than 250,-000 residents. Officials hiked the city's water rates in 1990, encouraging residents to install the package of home water-saving devices made available and to be more thrifty overall. And to bolster the whole effort, a large-scale public information campaign—including educating schoolchildren and airing radio and television spots—is under way to raise awareness about the city's water plight and to let people know how they can conserve.[6]

It is too early to judge the program's effectiveness, but officials are projecting that water use will fall from the current level of 300 liters per person per day to 250 by 1996, a drop of one sixth. Unfortunately, without a slowdown in birth rates and reduced migration to the capital, population growth will negate these per capita savings, and total water use in Mexico City will continue to climb, albeit at a slower pace.[7]

Although not in as dire straits as the Mexican capital, Waterloo in Canada has also shifted from the traditional approach of expanding supplies. A regional municipality with a population of 350,000 in the province of Ontario, Waterloo is Canada's largest metropolitan area dependent on groundwater for its supply. In the mid-seventies, signs of overpumping led officials to cap groundwater use and start looking for surface sources to meet new demands. The high price tags of the options considered—diversions from the Grand River and from Lake Huron some 120 kilometers to the west—led them to scale down their water development plans and to pursue conservation vigorously to cut the region's water demand.[8]

Through pricing, education, and the distribution of water-saving devices to make home plumbing fixtures more efficient, the Waterloo program has made conservation an effective part of its long-term water strategy. Volunteer groups have distributed retrofit kits (including toilet dams, faucet aerators, and low-flow showerheads) to nearly 50,000 homes, and homeowners have been urged to conserve water outdoors as well. Overall, Waterloo's per capita water use fell 10 percent in just three years.[9]

As in the case of Mexico City, Waterloo's efforts will be bolstered by province-wide efficiency standards for new plumbing fixtures, which will take effect in 1993. By 1996, new toilets throughout Ontario must meet a 6-liter standard, the strictest usually required today. Ontario has also set an ambitious goal of zero growth in water use for the next 20 years, which will get a boost from the conservation initiatives Waterloo already has under way. According to Ontario's Natural Resources Minister Bud Wildman, "If we achieve zero growth, we will reduce stress on the environment, lessen the likelihood of water shortages and reduce energy costs."[10]

For some cities, the high cost of treating wastewater is a prime motivator for conservation. In the mid-eighties, the sewage treatment plant in San Jose, California, was nearing capacity and the city was faced with building a new one at an estimated cost of $180 million. Since less water used indoors translates into less wastewater released to the sewer system, the city initiated a large-scale program to reduce residential and industrial water use quickly and reliably. The aim was to delay the need for this huge capital investment, saving the city and its residents money.[11]

The city set a goal in 1986 of cutting wastewater flows

to the treatment plant by 10 percent by 1996. The centerpiece of the conservation program was a massive retrofit campaign, in this case involving the distribution of water-saving devices door-to-door to some 220,000 households. Diligent canvassers made at least three attempts to talk with residents about the importance of installing the devices. As a result, 90 percent of the targeted households cooperated—a success rate unmatched by any other large-scale retrofit program. Water use in participating homes dropped 10–17 percent. With industry's savings added in, the conservation program had cut wastewater flows by an estimated 5.5 million cubic meters per year by 1991, one third of the goal.[12]

In the greater metropolitan area of Boston, Massachusetts, environmental concerns associated with a major expansion of the water supply forced a hard look at what conservation might achieve—and the result has been one of most comprehensive and successful programs in the United States. When demand in the metropolitan region rose above the reliable yield of the water supply system in the early seventies, water planners did what most do—looked for another source to dam or divert. In this case, engineers eyed the Connecticut and Merrimack rivers to the west, and proposed diverting some of their flows by tunnel to the metropolitan area.[13]

Environmental groups asserted that diverting the rivers would contaminate the city's otherwise relatively pure water supply, which did not require filtration, and also that the increased concentration of pollutants resulting from diminished river flows would damage salmon restoration efforts. They organized opposition to the projects in some 48 towns across the state, which, along with the engineering schemes' high price tags, led the city to seriously consider ways of curbing water demand.[14]

As a result, in March 1987, the Massachusetts Water Resources Authority (MWRA) launched an aggressive strategy of conservation and increased efficiency throughout its service area, which includes some 2.5 million people. Water-saving devices were installed in about 100,000 homes, leaks in old pipes were found and repaired, more than a million pieces of conservation literature were distributed to schoolchildren, and advice on water-saving measures was given to hundreds of businesses and industries.[15]

The results were impressive. Total annual water demand fell from 462 million cubic meters in 1987, when the program began, to 386 million cubic meters in 1991, a drop of more than 16 percent. (See Figure 12–1.)

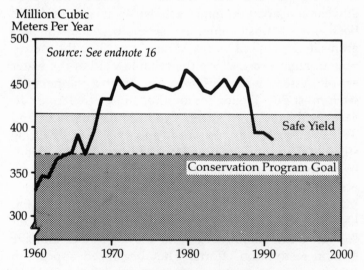

FIGURE 12-1. *Water Use in Greater Boston Metropolitan Area, 1960–91*

Water use is now below the system's safe yield. The MWRA plans to install water-saving devices in an additional 330,000 households during the next few years and to extend other aspects of the program as well. As a result of the expected savings, it has recommended postponing an expansion of the water supply until at least 1995. Moreover, the program has been cost-effective, with the conservation measures costing a third to half as much as the supply-side options that had been considered. MWRA's former Executive Director Paul Levy sums up the program's success: "For the first time in 20 years, we are living within our means."[16]

As these four cities illustrate, conservation makes sense for many reasons, and a different mix of measures will be best in different situations. In almost every case, however, successful efforts to curb domestic water use permanently will include some combination of economic incentives, regulations, and public outreach that together promote the use of water-saving technologies and behaviors. These measures are mutually reinforcing, and together they constitute a water supply option as reliable and predictable as a new dam and reservoir. As water becomes increasingly scarce, they often emerge as the least costly, most environmentally sound way of meeting community water needs when compared on an equal footing with traditional engineering approaches focused on expanding the supply.

Raising the price of water to better reflect its true cost is one of the most important steps any city can take. Proper pricing gives consumers an accurate signal about just how costly water is, and allows them to respond accordingly. Studies in a number of countries, including Australia, Canada, Israel, and the United States, suggest that household water use drops 3–7 percent with a

10-percent increase in water prices.[17]

Unfortunately, water is consistently undervalued, and as a result is chronically overused. Even worse, not only do water prices typically fail to promote efficiency, the water rate structures of many utilities actually reward waste by charging less per liter the more that is consumed. Seven out of ten residents in Manitoba, Canada, for instance, are charged according to this perverse "declining block" pricing policy, as are one out of three in Alberta and Ontario. Amazingly, water charges for most British households are linked to the value of their home, and have nothing to do with actual consumption.[18]

Many residences in both industrial and Third World cities are not equipped with water meters, which precludes even the possibility of charging people appropriately for their water use. Metering is not only a prerequisite to the success of most conservation measures, it encourages savings in and of itself simply by tieing the water bill to the amount used. In Alberta, the city of Edmonton meters all residential users, and its per capita water use is half that of Calgary, which is only partially metered. The areas of Calgary that are metered, however, register water use rates similar to Edmonton's. Trials in the United Kingdom have shown that metering can cut household use there by 10–15 percent.[19]

Raising water prices can often be politically difficult to do. But if accompanied by public outreach explaining the need for the price hike and the steps consumers can take to keep their water bills down, it can have a strong positive effect. When faced with dire water supply conditions in the mid-seventies, for instance, officials in Tucson, Arizona, raised water rates sharply to make them better reflect the true cost of service. At about the same time, they ran a public education campaign (called "Beat the Peak") with a goal of curbing water use on hot summer afternoons, when the supply was most in danger of running short of demand. The result was a 16-percent drop in per capita use within a few years, which, along with the lowered peak demand, allowed the Tucson water utility to cut its water supply expansion costs by $75 million.[20]

Pricing was the main tool of a conservation strategy adopted by the water utility serving Bogor, Indonesia, as well. With a proposed new water project estimated to cost twice as much per unit of water as existing supplies, the utility opted to try to reduce demand through more effective pricing. It tripled or quadrupled water prices, depending on the amount used, to encourage households to conserve. Between June 1988 and April 1989, average monthly residential water use dropped nearly 30 percent, which should allow the utility to connect more households to the urban water system at a lower cost.[21]

Since economic incentives and public outreach will not motivate everyone to conserve, setting water-efficiency standards for common fixtures—toilets, showerheads, and faucets—can be a critical component of a reliable conservation strategy. Standards establish technological norms that ensure a certain level of efficiency is built into new products and services. As already noted, Mexico has established nationwide standards, and Ontario, Canada, is including standards in its conservation strategy as well.

In the United States, there has been a growing movement among the states to mandate the use of water-efficient plumbing fixtures. In 1988, Massachusetts became the first to require that all new toilets installed use no more than 6 liters. Since then, 14 other states have followed suit, with most of them adopting efficiency standards for showerheads and faucets as well.[22]

Legislation that would set national standards has languished in the Congress for more than three years. Its passage would require that all new homes and major remodeling nationwide incorporate water-efficient fixtures and appliances. In this way, water savings would build over time, and they would be reliable and predictable. According to estimates by Boston-based water consultant Amy Vickers, the proposed standards would cause average U.S. indoor residential water use to fall gradually from 291 liters per person a day to 204, a 30-percent reduction. Prospects for passage of the standards brightened in May 1992, when, as part of a broader energy bill, the water efficiency provisions passed the House of Representatives and went to the Senate.[23]

Effective pricing, regulations, and public outreach can also help curb water use outdoors. In many dry regions, the sprinkling of lawns accounts for a third to half of residential water demand. This water has a particularly high economic and environmental price, since it is most needed during hot summer days when utilities experience their highest level of use. Meeting this "peak" demand requires planners to develop more water sources and treatment capacity than is needed to meet the amount of water typically used during most of the year.

A water-saving practice many communities in the United States have turned to in recent years is "Xeriscape landscaping." From the Greek word xeros, meaning dry, Xeriscape designs draw on a wide variety of attractive indigenous and drought-tolerant plants, shrubs, and ground cover to replace the thirsty green lawns found in most suburbs. A Xeriscape yard typically requires 30–80 percent less water than a conventional one, and can reduce fertilizer and herbicide use as well. One study in Novato, California, found that Xeriscape landscaping cut water use by 54 percent, fertilizer use by 61 percent, and herbicide use by 22 percent.[24]

Just a decade old, the Xeriscape concept has spread rapidly through parts of the United States. Programs in at least eight states (including several in the more humid East) actively support Xeriscape landscaping as a way of conserving water and improving the urban environment. Tucson, Arizona, gave it a boost in early 1991, for instance, by forbidding new developments from having more than 10 percent of their landscape area planted in grass. The practice is making inroads in a handful of other countries as well, such as Australia, Canada, and Mexico.[25]

Besides cutting indoor and outdoor water use, a comprehensive urban conservation effort will also curb water losses in the water distribution system itself. Especially in older cities, finding and repairing leaks usually yields a big payoff. As urban water systems deteriorate because of age or lack of maintenance, large amounts of water can be lost through broken pipes and faults in the distribution network. More than half the urban water supply simply disappears in Cairo, Jakarta, Lagos, Lima, and Mexico City. Although some of this water is probably siphoned off by poor residents not served by the system, much of it gets to no one. Moreover, these are costly losses because this "unaccounted-for water" is collected, stored, treated, and distributed, but never reaches a billable customer.[26]

In most cases, finding and fixing leaks rewards a city

not only with water savings, but with a quick payback on the investment. At a cost of $2.1 million, the MWRA's leak detection program cut system-wide demand in the greater Boston area by about 10 percent, and was one of the most cost-effective measures in its whole strategy. Leak detection and repair can be especially beneficial in developing-country cities with extremely large losses, since the existing water supply system can then serve many more people. Reducing Jakarta's "unaccounted-for water" from 51 percent to 31 percent, for example, would retrieve some 45 million cubic meters annually, enough to supply 800,000 people.[27]

With some notable exceptions, such as Mexico City and Bogor, Indonesia, few Third World cities are actively trying to conserve water. Most are preoccupied with the daunting challenge of providing reliable water services to the large number of people now lacking them. Given that average household use in most developing countries is a fraction of that in industrial countries, conservation and efficiency are often viewed as irrelevant or, at best, options to pursue later.

Quite to the contrary, conservation is an integral part of any practical solution to the water supply problems of poorer nations. With the Third World's population growing by 90 million each year and with widespread migration from rural areas to cities, the stage is set for explosive urban growth. Water-efficient hardware, pricing policies, and other measures offer an opportunity for these cities to build conservation into their water plans from the start, allowing more needs to be met with fewer resources and curbing water costs overall.[28]

Indeed, for the developing world to adopt the water-intensive ways of industrial countries—which are now themselves finding their water practices unsustainable—

would be a costly mistake. Constructing water distribution networks, connecting each individual household to water and sewer pipes, and building centralized water and wastewater treatment plants costs in the range of $450–700 per person served. By holding down each household's water demand, water-efficient plumbing fixtures and other conservation measures can help lower these costs. They allow expensive new treatment plants and distribution pipes to be scaled down in size, reducing both capital and operating expenditures.[29]

There are now some positive signs that the idea of incorporating conservation into long-term water supply planning in the developing world may be catching on. The World Bank, in collaboration with the United Nations Development Programme, has started to work with a number of countries, including Chile, China, India, and South Korea, to identify cities that could serve as useful demonstration sites for urban conservation. Showing and spreading the word about conservation's benefits could help more cities in developing countries meet the water needs of their residents while avoiding many of the excesses that so many urban areas in the industrial world are now trying to eliminate.[30]

III

Toward
Water Security

13

Pricing, Markets, and Regulations

Together, the many ways of conserving, recycling, and reusing water described in Part II constitute the makings of an efficiency revolution. With tools and technologies readily available, enormous water savings are possible in agriculture, industries, and cities. Yet we are stuck at the brink of this transformation because of policies and laws that encourage wastefulness and misuse rather than efficiency and conservation.

Moving toward more efficient, ecologically sound, sustainable patterns of water use requires major changes in the way water is valued, allocated, and managed. Appropriate pricing, the creation of markets for buying and selling water, and other economic inducements for wise water use hardly exist in most places. They have a central role to play in the transition to an era of scarcity. In addition, protecting the many functions water performs that a marketplace does not adequately value—such as habitat protection, species preservation, recreational uses, and aesthetic benefits—requires limiting the amount of water that cities, industries, and agriculture collectively claim. And finally, with the stability of the water cycle so dependent on the land over which water flows, regulations on how we use critical parcels of the earth are also necessary to achieve water security.

Many of the water shortages cropping up around the world stem from the widespread failure to value water at anything close to its true worth. Grossly underpricing water perpetuates the illusion that it is plentiful, and that nothing is sacrificed by wasteful practices. Benjamin Franklin once said, "When the well's dry, we know the worth of water." A key challenge is to begin valuing it appropriately and using it more wisely so as to avoid learning Franklin's lesson the hard way.[1]

Setting prices closer to the real cost of supplying water is a key component of both urban and industrial conservation. As noted in earlier chapters, this would encourage both city-dwellers and manufacturers to conserve, recycle, and reuse their supplies, thereby fostering greater efficiency among those accounting for a third of the world's total water use.

Pricing water properly is most important in agriculture, however, because wasteful irrigation constitutes the single largest reservoir within the "last oasis." Water subsidies are larger and more pervasive in agriculture than in any other realm of water use. Governments often build, maintain, and operate irrigation systems with public funds, and then charge farmers next to nothing for these expensive services. Irrigators in Mexico, for instance, pay on average just 11 percent of their water's full cost, and those in Indonesia and Pakistan, about 13 percent. In Egypt, a land of extreme scarcity, farmers are not charged directly for their irrigation water at all.[2]

In India, the world's third largest food producer, government spending to operate and maintain medium and large canal projects exceeds the total revenue collected from farmers by 23.5 billion rupees ($816 million). Adding in capital cost subsidies would lift this figure even higher. Irrigation officials set water charges according to the size of farmers' plots and the crops they are growing, so payments bear no relation to the amount actually used on the fields. Moreover, charges are so low—typically amounting to 2–5 percent of the harvest's value—that they have no influence on farmers' management decisions. Water prices have not been raised in most Indian states since the mid-eighties and in a few, including the Punjab and water-short Tamil Nadu, since the mid-seventies.[3]

Such extreme undercharging not only fosters waste and the planting of water-intensive crops, it also deprives government agencies of the funds needed to maintain canals and other irrigation works adequately. As a result, agriculture usurps far more water than is necessary for the harvest it yields, farmers grow thirsty crops like sugarcane even in water-tight areas, and irrigation works fall into disrepair, which reduces efficiency even further.

The situation is little better in the United States. The federal Bureau of Reclamation supplies water to a quarter of the West's irrigated land—more than 4 million hectares—under long-term (typically 40-year) contracts at greatly subsidized prices. This practice dates back to the 1902 Reclamation Act, which aimed to settle the western frontier by helping family farmers obtain irrigation water and power. Over time, the degree of federal assistance deepened with the bureau's decision not to charge interest on water project construction costs, to lengthen the repayment period, and to limit repayment to farmers' "ability to pay."[4]

As a result, subsidies ballooned over time, with farmers large and small drinking heavily at the public trough for decades. As of the mid-eighties, irrigators benefiting from California's huge Central Valley Project had repaid only 4 percent of its capital costs: $38 million out of $950 million. U.S. taxpayers have footed the bill for the remainder.[5]

As in poorer countries, this free ride has discouraged farmers from investing in efficiency improvements and has led them to choose crops ill-suited to a semidesert and to devote scarce water to low-value uses. A third of the water delivered by the Bureau of Reclamation is used to irrigate hay, pasture, and other forage crops destined for livestock. Meanwhile, western cities and industries scramble for more water, and drum up plans to dam yet another canyon or divert even more from a distant river.[6]

Correcting these perverse situations is easier said than done. It requires bucking deeply entrenched and politically influential special interests, instilling irrigation bureaucracies with a broader sense of mission, and decentralizing water management so that local water suppliers and users have more responsibility and accountability for the performance of their operations. In some cases, it even means challenging religious and cultural beliefs. Islamic norms, for instance, hold that water must be free, which has generally precluded governments in Muslim countries from charging anything more than the cost of delivery.[7]

Requiring farmers in developing countries to at least pay for the operation and maintenance of their irrigation systems is often frustrated by the notion that they cannot afford higher prices. Yet those benefiting from irrigation typically earn far more than those cultivating rainfed lands do. Lessening irrigation subsidies would free up funds to invest in the productivity of rainfed farming, which accounts for the bulk of the world's cropland and provides the livelihood of most of the rural poor. Moreover, Third World irrigators have shown time and again that they are willing and able to pay more for water that is reliable and over which they can exercise control. With an assured and timely supply, they can invest in fertilizers, high-yielding seeds, and better management practices, often boosting their crop production and income enough to offset any rise in water prices.[8]

Reducing irrigation subsidies would thus tend to promote both efficiency and equity while stemming problems of waterlogging, soil salinization, and other forms of environmental degradation. Precisely how this is done will vary from case to case, and will not always be easy. The canal systems in the Third World often span huge areas; some in India, for instance, cover more than a million hectares and include thousands of farmers. Measuring the exact amount of water each individual irrigator uses and charging for it appropriately would be a costly administrative nightmare.

But practical ways do exist to give farmers an economic incentive to use water more efficiently. In a pilot project in Maharashtra, India, for example, a local nongovernmental organization helped a farmers association and the irrigation bureaucracy craft a workable arrangement for charging farmers according to the volume of water taken from the minor irrigation canal that services their cropland area. It is up to the group to determine how much each individual farmer will pay. Any water they are entitled to but do not use is stored in a reservoir and supplied to them during the dry season, when they otherwise might not get any water at all. This gives the farmers association a powerful motivation to use water sparingly, both because they pay for the amount they take and because their efficiency determines how much they get during the dry season, when irrigation is so critical to the harvest.[9]

Such a scheme shows that with creativity and flexibility, incentives to promote efficiency can be designed. A key in almost all cases is ensuring more local accountability for how irrigation systems perform. At a minimum, setting water fees to cover operation and maintenance costs, collecting them through a local farmers organization that is involved in the system's management, and making irrigation managers accountable for the performance of their project areas could go a long way toward realizing irrigation's potential in the Third World.

Since farmers relying on groundwater typically operate their own wells, direct water subsidies to them are much lower. But many Third World governments greatly undercharge for energy, and, since pumping is a large share of total groundwater costs, this effectively amounts to a large irrigation subsidy. India's rural electricity subsidies, for instance, totaled some 14.6 billion rupees ($507 million) per year in the mid-eighties, contributing to the pervasive and worsening problem of falling water tables. Where groundwater is used for irrigation, eliminating energy subsidies can thus be as critical to water conservation as charging more for water is.[10]

In the United States, meaningful reform of the Bureau of Reclamation's irrigation policies has repeatedly fallen victim to the powerful western agricultural lobby and the politicians beholden to them. With hundreds of federal irrigation contracts coming up for renewal this decade, a timely opportunity exists to establish new rules for what has become a very different water game. A first step is to reduce the subsidies that give farmers water for a small fraction of the price cities and industries pay. Prices could be raised gradually, say over five years, to avoid sudden disruptions. Once farmers know that water prices will be much higher in the next round of contracts, the government could induce them to renegotiate sooner by offering to help pay for conservation investments undertaken immediately.[11]

Unfortunately, Department of Interior officials have shown little interest in substantially raising prices, seriously redressing irrigation's environmental damage, or making other needed changes. As of late June 1992, legislation is under debate in Congress to reform California's large Central Valley Project, which could set the stage for broader reform of western irrigation systems. Meanwhile, the waste and ecological damage persist— all for the benefit of a few and at taxpayer expense.[12]

Try as it might to maintain the status quo, agriculture cannot lock up its customarily large share of regional water supplies for long. As most schoolchildren learn, when the pie stops getting bigger, some must eat less if others are to eat more. With the pace of water development slowing and supplies no longer expanding in places, meeting new demands will increasingly require shifting water among the different users—irrigators, industries, cities, and the natural environment.

In many parts of the world, such competition is already evident, and in most cases it is agriculture that will lose water—sometimes out of choice, and sometimes not. In north China, reservoirs originally built to serve farmers now supply growing urban areas and industries, which yield more economic value per unit of water than crop production does. Israeli officials plan to shift more than a third of agriculture's fresh water over to cities during the coming decades. Water constraints around New Delhi, Madras, and other Indian cities are breeding rivalries there as well.[13]

In the western United States, competition for scarce supplies has spawned an active market that is fostering transfers of water from farms to cities. Where farmers have clear property rights to water, as they do in much of the West, they have the option of selling those rights to a willing buyer. If an irrigator can earn more by selling water to a nearby city than by spreading it on alfalfa, cotton, or wheat, transferring that water from farm to city use is economically beneficial. If it prevents the city from damming another river to increase its supplies, the transfer can also benefit the environment. In this way, marketing can be an effective means of reallocating a finite pool of water.

Farmers can free up supplies for sale in three ways—by irrigating more efficiently and selling the conserved water, by switching to less thirsty crops and selling the water they no longer need, or by taking land out of irrigation entirely and either producing dryland crops or retiring the land from agriculture. Irrigators may also choose among several different types of transactions. For example, they can sell their water rights directly, which permanently transfers control to the buyer. They can lease some or all of their water for an agreed-upon period, while keeping the rights. Or they can swap supplies with another water user.

During 1991, 127 water transactions of various kinds were reported in 12 western states, up slightly from the 121 reported in 1990. Almost all the water sold or leased in 1991 came from irrigation, and two thirds of the trades resulted in cities getting more water for immediate or future use. Prices varied greatly. In Colorado, where half the transactions took place, water traded for about $2,140 per acre-foot ($1.74 per cubic meter), nearly twice the average trading price in 1989. Much of the price run-up is attributed to the Environmental Protection Agency's veto of the Two Forks Dam, which would have increased supplies for the Denver area.[14]

Exactly how far U.S. water trading ultimately will go in reallocating supplies remains unclear. According to some estimates, redirecting 7 percent of western agriculture's water to cities would be sufficient to meet the growth in urban demand projected for the end of the decade. After that, larger shifts would be needed. Unless cities stabilize their water use through conservation, reuse, and, where necessary, limits on the size of their populations and economies, agriculture ultimately could lose more water—and land—than is socially desirable, given the challenge that lies ahead of feeding a much larger world population.[15]

To the extent that agricultural supplies are freed up by increasing irrigation efficiency or by switching crops, land need not come out of production. For instance, the Metropolitan Water District (MWD) of Southern California—water wholesaler for roughly half the state's 30 million residents—is financing the lining of canals and other conservation projects in the neighboring Imperial

Irrigation District in exchange for the 100,000 acre-feet of water per year the investments will save. The annual cost per acre-foot conserved is estimated at $128, far lower than MWD's best new-supply option. Enough water is being traded this way to meet the annual needs of 200,000 households, yet no cropland is being taken out of production. Since the irrigation district keeps control of the water rights, however, MWD is only assured of these supplies for the 35 years covered by the agreement.[16]

Transactions in Arizona, on the other hand, have caused great controversy, since Phoenix, Tucson, and other rapidly expanding cities have taken to "water ranching." State law makes it difficult to buy rights to water independent of the land, so cities have purchased farmland for the water that comes with it. This elimination of farming has threatened to deprive rural communities of tax revenue and income needed to keep them viable. Passage of a new state law in 1991 limits future farm water exports to land already acquired by cities, along with two other groundwater basins. It also requires cities to pay a sum equal to the property taxes that would have been assessed on those farms had they not been sold. So while irrigated agriculture will continue to shrink in Arizona, it will likely do so to a lesser degree and at a slower pace as a result of the new law.[17]

In parts of Bangladesh, India, and Pakistan, marketing has emerged as an effective way of distributing water more equitably, particularly in areas where irrigation relies on groundwater. Often the poorest villagers cannot afford the pumps and other machinery needed to extract underground water for their crops. But if they are able to buy supplies from wealthier farmers, they can still receive some of irrigation's benefits—including higher and more reliable yields, and more certain income. The resulting expansion of irrigation would also create more stable employment opportunities for landless people in the vicinity, since more dry-season cropping would occur.

In one irrigated area of Bangladesh, each owner of a shallow groundwater well sells water to an average of 14 other farmers. For each hectare irrigated by a well owner there, two other hectares are irrigated by farmers purchasing water. Indian agricultural economist Tushaar Shah notes that evidence from a number of Indian states, including Andhra Pradesh, Gujarat, and Tamil Nadu, suggests that where a poorer farmer can buy enough water to grow an additional crop, water selling "can have dramatic beneficial impacts on the incomes of water buyers and the economy of the community as a whole."[18]

In some cases, buyers pay cash for the water, but often they pay in kind, either by lending their labor, sharing a portion of their harvest with the seller, or some combination of these two approaches. Where electricity is priced according to a flat fee geared to the horsepower of the pump, as is common in many areas, a farmer has an extra incentive to sell water, since there is no real cost to the extra pumping. Although this makes relatively inexpensive water available to poorer farmers, it also creates a strong inducement to overpump the resource, especially because tubewell owners often have de facto ownership rights to as much groundwater as they can extract. To serve the goals of efficiency, equity, and sustainable resource use simultaneously, water marketing would need to be accompanied by limits on groundwater pumping, the reduction of energy subsidies, and assurances that markets do not further concentrate water rights in the hands of the rich.[19]

Indeed, wherever pricing and marketing fail to take into account the full social, environmental, and inter-generational costs of water use, some additional correction is needed. In areas with declining groundwater levels, for instance, governments can limit the total amount pumped to the average rate of aquifer recharge. In the United States, Arizona pioneered this approach in 1980 with passage of a law requiring that groundwater basins undergoing depletion achieve a balance between pumping and recharge by the year 2025. Unfortunately, many cities affected by this legislation have not met their conservation targets, and, as noted earlier, have sought to buy land and water rights from farms.[20]

Another option is to tax groundwater pumping that exceeds natural replenishment. A 1991 Arizona law moves toward this approach in the Phoenix area, requiring those who have overdrawn their groundwater accounts to pay a "replenishment tax" or to purchase credits from someone who has pumped less than the allowable level. The tax rate reflects the cost of supplying enough water to balance the whole district's account, and is thus high enough to induce conservation.[21]

In the case of fossil aquifers, such as the Ogallala in the U.S. High Plains or the deep desert aquifers in Saudi Arabia and Libya, this approach could take the form of a "depletion tax" on all groundwater extractions. In this way, those profiting from the draining of one-time reserves would at least partially compensate society for the loss of these supplies, which will be valued far more highly by future generations faced with feeding a much larger world population. A groundwater depletion tax would help promote equity among generations—a basic tenet of a sustainable society—while also helping slow the depletion rate by encouraging conservation.

Public action is also needed to ensure that ecological systems get the water they need to remain healthy. Open markets that allow water to be purchased and dedicated to this purpose can help. For instance, 11 of the 127 water transactions in the western United States in 1991 were aimed at securing more water for rivers, wetlands, and nature preserves.[22]

But the task at hand is far larger than private conservation initiatives alone can handle through the marketplace. According to Defenders of Wildlife, a U.S. conservation group, in 1989 (a drought year) wildlife refuges in California's Central Valley got less than 8 percent of the water needed in order for migrating waterfowl to winter there successfully. Private action simply cannot secure the large volumes of water needed to serve the public's interest. Collecting money from the millions of people who are willing to pay for protection of these ecological values is far too difficult and costly. Economists call this a problem of excessively high "transaction costs," and it is an important reason the market fails to protect the environment adequately.[23]

In such instances, regulations are needed to preserve and restore ecological health. The water laws and practices of most countries are heavily biased toward the individual's right to withdraw water for private gain and against the public's common interest in leaving water "instream" to maintain fisheries, recreational values, and the integrity of ecosystems. Where water is plentiful, the consequences of this bias may be negligible; but where it is scarce, severe ecological damage results, as is now so evident in many parts of the world.

One way of protecting water's life-support functions is simply to limit the total amount that can be diverted from a river, lake, or stream. Until fairly recently, this was difficult in the western United States, since water rights had to be put to "beneficial use," which was interpreted as removing water from its natural channel for some productive purpose. Most states, however, now recognize water left "instream" to protect ecological functions as a legitimate beneficial use to which water rights can be attached. Only a few allow individuals and private entities to hold instream rights; in most cases, a state agency must acquire them. Montana, for instance, passed a law in 1973 that allows the state and federal governments to reserve water for instream uses. As a result, about 70 percent of the average annual flow in the upper basin of the Yellowstone River and half to two thirds of the lower basin flow have been reserved to protect aquatic life, water quality, and other ecological services.[24]

Where excessive diversions have already caused ecological damage, as with central Asia's Aral Sea or Florida's Everglades, new laws and regulations will be needed to restore ecosystems to health. One such instrument is a legal principle called the "public trust doctrine," which asserts that governments hold certain rights in trust for the public and can take action to protect those rights from private interests. Widespread application of this doctrine could have sweeping effects, since even existing water rights could be revoked in order to prevent violation of the public trust.

In a landmark decision handed down in February 1983, the California Supreme Court declared that the water rights of Los Angeles, which permit diversions from the Mono Lake basin, are subject to the public trust doctrine. Mono Lake—a hauntingly beautiful water body on the eastern side of the Sierra Nevada whose algae and brine shrimp support hundreds of migratory bird species—has had its volume halved and its salinity doubled because of excessive diversions from its major tributaries. Since 1989, the courts have prevented the Los Angeles Department of Water and Power from siphoning off any Mono basin water, which previously had constituted some 15 percent of the urban utility's supply. With a final determination of the city's rights to Mono water not expected until 1993, Judge Terrence Finney called Mono Lake "a national environmental, ecological and scenic treasure [that] should not be experimented with even for a few brief years."[25]

Protecting water systems also depends on regulating the use of those critical areas of land that help moderate water's cycling through the environment. Degradation of the watershed—the sloping land that collects, directs, and controls the flow of rainwater in a river basin—is a pervasive problem in rich and poor countries alike. Besides contributing to flash floods and loss of groundwater recharge, which can exacerbate the effects of drought, it leads to soil erosion that prematurely fills downstream reservoirs with silt, shortening the useful life of these expensive water projects.

When the Ubolratana dam in Thailand was completed in 1965, for instance, 90 percent of the upper catchment area was forested. Less than two decades later, forest cover was down to 40 percent, ironically in part because people living on the reservoir site were relocated there. Erosion rates sharply increased, greatly reducing the useful storage area of the reservoir. Worldwide, the replacement cost of reservoir capacity lost to siltation is estimated to total about $6 billion a year.[26]

With today's population pressures and the need for greater food production, keeping entire watersheds forested is no longer possible in most places. About half of Asia, for example, is technically watershed, defined as land sloping at an angle of at least 8 degrees. Much of this land is, and needs to be, in agriculture. Yet in sensitive regions, governments may now need to require cultivation practices that protect basic watershed functions, especially in the steeply sloping upper basin areas.[27]

Fortunately, many of the measures that can help safeguard water supplies also enhance crop production in upland areas. Terracing, mulching, agroforestry (the combined production of crops and trees), and planting vegetative barriers on the contour are just a few of the ways soil and water can be conserved while improving agricultural output. Cultivating on the contour, for instance, on slopes up to 30 degrees has produced yields 6–66 percent higher than traditional cultivation up and down the slope. On lands not suitable for cultivation, revegetating deforested slopes, reducing grazing pressures, and altering timber practices are among the menu of options for watershed protection. The challenge for local and national governments is to plan the use of watershed lands with soil and water conservation in mind, recognizing that the way uplands are managed greatly affects the livelihoods of people and the integrity of water systems downstream.[28]

Land use planning in and around cities and suburbs can be equally important for the protection of local water supplies. Unplanned development can end up paving over rainwater's main point of entry into a key drinking-water source. Especially in areas dependent on local groundwater, protecting these critical aquifer recharge areas is essential to ensure that water sources get replenished. Suffolk County, Long Island, recently spent $118 million to acquire 3,440 hectares of open space in order to preclude development in recharge zones vital to the region's underground water supply, its sole source of drinking water. Voters approved a one-quarter cent increase in the county sales tax to fund the land purchases, which remains in effect until the end of the decade.[29]

Local ordinances can also set landscaping requirements with an eye toward protecting water supplies. Across the United States, cultivated lawns cover 10–12 million hectares, an area about the size of Kentucky. Not only do lawns fail to promote recharge effectively in many cases, the fertilizers and pesticides used to maintain them are troubling sources of pollution. The town of Southampton, Long Island, requires that at least 80 percent of each home lot situated in a critical aquifer zone be kept in its natural (typically wooded) state and that no more than 15 percent be put in lawns or vegetation that require fertilizer.[30]

A number of states—including Connecticut, Georgia, New York, and North Carolina—have adopted laws and regulations specifically designed to control land use in watersheds. North Carolina passed a law in 1989, for instance, requiring the development of minimum statewide standards for watershed protection, which were due out by July 1992. Cities and towns are required to develop land use plans and ordinances that are at least as strict as the state's standards, which include, for example, limits on impervious surface areas and certain agricultural practices.[31]

Here and there, pricing, marketing, and regulatory actions are being used effectively to tap the "last oasis" of conservation and efficiency and to promote sustainable water use. But nowhere have all the elements been brought together into a strategy ensuring that human use of water remains within ecological bounds and that the integrity of water systems overall is protected. We can muddle through by fine-tuning antiquated policies and practices, but only for so long. And the longer we wait to make the needed adjustments, the costlier they are likely to be—with a greater loss of ecological assets in the interim.

14

A Water
Ethic

Why has so much of modern water management gone
awry? And why is it that ever greater amounts of money
and ever more sophisticated engineering have not solved
the world's water problems?

The answers, in part, lie in the overarching message of
this book: we are trying to meet insatiable demands by
continuously expanding a supply that has limits, both
ecological and economic. The host of nuts-and-bolts
measures described in Part II—from thrifty irrigation
techniques and rainwater harvesting to water-saving
plumbing fixtures and wastewater recycling—can help
us out of this predicament by reducing the amount of
water required to grow food, produce material goods,
and meet household needs. As described in Chapter 13,
these measures will only spread rapidly if economic in-

centives and regulations are adopted that promote conservation and efficiency instead of waste and profligacy.

Yet there is something missing from this prescription,
something less tangible than drip irrigation lines and
low-flow showerheads—but in the final analysis as important. For at the heart of the matter is modern society's disconnection from water's life-giving qualities.
For many of us, water simply flows from a faucet, and
we think little about it beyond this immediate point of
contact. We have lost a sense of respect for the wild
river, for the complex workings of a wetland, for the
intricate web of life that water supports. By and large,
water has become strictly a resource to be dammed, diverted, and drained for human consumption.

Grasping the connection between our own destiny
and that of the water world around us is integral to the
challenge of meeting human needs while protecting the
ecological functions that all life depends on. Our farms,
factories, and homes are not just competitors for a resource, they are members of a community embraced
and supported by the ecosystems around them. To
manage water as if it were separate and apart from us is
like cutting off the flow of blood to one part of the body
in order to send it to another—the living entity suffers,
and, depending on where the diversion takes place, may
not survive.

We have been quick to assume rights to use water, but
slow to recognize obligations to preserve and protect it.
For sure, better pricing and more open markets will as-
sign water a higher value in its economic functions, and
breed healthy competition that weeds out wasteful and
unproductive uses. But we also need a set of guidelines
and responsibilities that stops us from chipping away at
natural systems until nothing is left of their life-sustain-

ing functions, which the marketplace fails to value adequately. In short, we need a water ethic—a guide to right conduct in the face of complex decisions about natural systems we do not and cannot fully understand.[1]

The essence of such an ethic is to make the protection of water ecosystems a central goal in all that we do. This may sound like an idealistic prescription in light of the ever more crowded world of needs and aspirations in which we live. Yet it is no more radical a notion than suggesting that a building be given a solid foundation before raising it 10 stories high. Water is the basis of life, and our stewardship of it will determine not only the quality but the staying power of human societies.

Adopting such an ethic would represent a historic philosophical shift away from the strictly utilitarian, divide-and-conquer approach to water management and toward an integrated, holistic approach that views people and water as related parts of a greater whole. It would make us stop asking how we can further manipulate rivers, lakes, and streams to meet our insatiable demands, and instead to ask how we can best satisfy human needs while accommodating the ecological requirements of healthy water systems. And it would inevitably lead us to deeper questions of human values—in particular, how to narrow the unacceptably wide gap between the haves and have-nots while remaining within the bounds of what natural systems can sustain.

Living by such an ethic would mean using less whenever we can, and sharing what we have. It is about being good neighbors—as individuals, companies, communities, states, provinces, and nations. And it establishes norms of responsible behavior against which to judge the actions of each global citizen.

In practical terms, a water ethic is part of a sustainable

development code that entails a wholly new approach to economic progress, one that harmonizes economic goals with ecological criteria. As shown in earlier chapters, the single-minded pursuit of crop production, industrial growth, and urban expansion has damaged numerous aquatic ecosystems that support fisheries, provide habitat for waterfowl and other wildlife, and protect water quality. Besides causing tangible monetary losses, this deterioration signals that systems essential to our own well-being are breaking down and that environmental services that we take for granted are being lost.

A society guided by a water ethic would have a set of indicators to monitor these trends, and would make course corrections to restore ecosystems to health before irreparable harm is done. We may see glimmerings of such an ethic at work where scientists are trying to undo the damage caused by unsustainable economic development, as in south Florida's Everglades, California's Kesterson wetlands, and central Asia's Aral Sea basin. But these are attempts to bring ecosystems back from the brink of destruction. They are costly, and offer no guarantees of success. Were an ethic operating at the outset, different economic choices would have been made, and the degree of violence done to nature greatly lessened.

For a test case of ecologically guided development, we might watch Botswana's Okavango Delta. (See Chapter 5.) All the usual pressures and tensions to develop water for economic gain are there, threatening to ravage this unique wild place. But so far, the government has listened to local people who rely on the delta's abundance of fish and wildlife for their livelihood, and who want the wetland to remain intact, and it has heeded studies that show a proposed river diversion project is not really

needed. All too rare, such ecological sensitivity and respect for democratic traditions are a good part of what
sustainable development is all about.

Governments, the World Bank, and other lending
institutions that help set water priorities and directly
fund many projects can help make this new ethic operational by placing ecological sustainability at the core of
their investment policies and decisions. More often
than not this will favor smaller-scale projects—such as
pumps and wells for groundwater development, micro-
catchments, or small reservoirs to collect and store local
runoff—over large dams and river diversions that tend
toward greater disruption of natural systems. Commu-
nity-based projects are also better able to focus on meeting the needs of the poorest people and to engage local
people in development efforts from the outset—
prerequisites for sustainable economic progress.

Striving to do more with less would also place options
to reduce water demand on equal footing with conventional engineering projects aimed at continuously expanding human access to nature's supply. Despite a
wealth of evidence showing conservation and efficiency
to be among the most economical and environmentally
sound alternatives, these are often still perceived as only
minor additions to the water supply package. Installing
low-flush toilets, lining irrigation canals, and recycling
factory water lack the fanfare and political appeal of a
big new dam. But they are at the core of cost-effective
and sustainable solutions to achieving water balance.

In many cases, a water ethic would require a reorder-
ing of economic goals and priorities. As water becomes
scarce, sustainable development depends on raising its
productivity—getting more value out of each liter
used—while leaving enough in rivers, lakes, and aquifers

to keep natural systems functioning well. This may mean, for instance, that boosting food self-sufficiency is not a wise or realistic goal for some countries.

Egypt, for example, will soon exceed its water budget, yet the government is attempting to reclaim an additional 60,000 hectares of desert each year in order to expand crop production. Given the projected growth in population and today's level of per capita water use, the nation's total water demand 20 years from now will exceed its allotted share of the Nile by nearly 60 percent. Long before this point, ecosystem functions will be grossly impaired as streamflows diminish and water quality deteriorates. Although conservation and reuse can trim the water deficit and slow the pace of ecological damage, there is no getting around the need to reduce irrigated agriculture in order to buy time to slow population growth.[2]

The "good neighbor" aspects of a water ethic come into play at all levels of human interaction—from personal behaviors and life-style choices to international relations and commitments. It remains a grave moral shortcoming that 1.2 billion people cannot drink water without risking disease or death. The reason is not so much a scarcity of water or inadequate technologies as a lack of social and political commitment to meeting the basic needs of the poor. It would take an estimated $36 billion more per year, equal to roughly 4 percent of the world's military expenditures, to bring to all of humanity what most of us now take for granted—clean drinking water and a sanitary means of waste disposal.[3]

Stronger notions of equity and fairness must also be interjected into international relations if tensions over water are to be defused. Like oil, water is a strategic resource for which nations will compete fiercely as it becomes more scarce. International law embodies an ethic of equity among nations within a watershed or river basin, and lays out basic notions of rights and responsibilities with regard to shared watercourses. But clearer criteria are needed by which to judge, for instance, what constitutes a reasonable level of per capita use given the total amount of water available in a river system, and what constitutes a fair apportioning of water among nations sharing common sources.

In the meantime, it is up to neighboring countries to work out mutually agreeable water-sharing terms and treaties. Averting outright water wars in the Middle East will take all the creativity and cooperation countries in this region can muster. Recognition that peace without water security can be a tenuous truce at best has at least brought some nations to the bargaining table. Once there, the task is to turn what is often perceived as a zero-sum predicament—in which one party's gain is another's loss—into a win-win proposition.

By cooperating to expand the available supply through conservation, efficiency, and reuse, nations in the Middle East can ease water tensions, enhance prospects for peace, and better protect their environments. A collaborative effort, perhaps with international funding, to help Jordan cut its water use per irrigated hectare by a third—as Israeli farmers have done—could save some 170 million cubic meters per year, equal to half the natural recharge of the West Bank aquifer. If, in return for technical assistance, Israel received a portion of the saved water, all parties might benefit and tensions over water would be lessened.[4]

For individuals, a water ethic calls for an examination of life-styles and consumption patterns with an eye toward reducing personal claims on the earth's finite water supply. As an essential ingredient in most manufacturing operations and the receptacle for much of our waste, water has a role in virtually every product we buy and is polluted by much that we throw away. We rarely think about water when we see an automobile, for example, but producing a typical U.S. car requires more than 50 times its weight in water.[5]

Especially for those 1 billion of us in the high-consumption class, cutting down on our purchases of material things—from clothes and shoes to paper and appliances—conserves and protects water supplies as effectively as installing a low-flush toilet does. As with so many natural resources, as long as prices in the marketplace fail to reflect full social and ecological costs, voluntary changes in consumption patterns will play an important role in the quest for sustainability.

Calls for life-style changes may start as an invitation to the ethically minded, but before long may progress to an ecological imperative. For example, 38 percent of the global grain harvest today becomes feed for livestock. Especially where feedgrain is irrigated, animal agriculture is highly water-intensive. A kilogram of hamburger or steak produced by a typical California beef cattle operation, for instance, uses some 20,500 liters of water. As the world struggles to grow enough food for a larger population, the amount of water needed for feedlot-style meat production may preclude meat-rich diets. By shifting diets away from meat and toward more grains and legumes, consumers in the more affluent countries can get their protein requirements in a way less demanding of the earth's endowment of fertile land and fresh water.[6]

Ultimately, securing sufficient water for people while leaving enough for a healthy environment overall depends on a rapid slowing of population growth. If current trends persist, per capita water supplies worldwide will drop by more than a third by 2025. Efforts to raise living standards for the additional 3.1 billion people alive then will place enormous strains on natural systems. Moreover, human numbers continue to grow fastest in some of the most water-short regions. At current growth rates, the populations of 18 of the 20 countries in Africa and the Middle East now qualifying as water-scarce will double within 30 years. No set of technological feats, however imaginative, can win such a race. Reducing birth rates through comprehensive family planning and equal economic opportunity for women— vital for so many other reasons—is essential to a secure water future.[7]

Falling freely from the sky, water has deluded us into believing it is abundant, inexhaustible, and immune to harm. The challenge now is to put as much human ingenuity into learning to live in balance with water as we have put into controlling and manipulating it. The "last oasis" of conservation, efficiency, recycling, and reuse is large enough to get us through many of the shortages on the horizon, buying us time to develop a new relationship with water systems and to bring consumption and population growth down to sustainable levels.

Yet the pace of this transition needs to quicken if we are to avert severe ecological damage, economic setbacks, food shortages, and international conflicts. In the end, the time available to adjust may prove as precious as water itself.

Notes

CHAPTER 1. An Illusion of Plenty

1. Annual precipitation from George H. Hargraves, *World Water for Agriculture* (Logan: Utah State University, 1977).
2. I.A. Shiklomanov, "Global Water Resources," *Nature & Resources*, Vol. 26, No. 3, 1990.
3. Industry's use from Shiklomanov, "Global Water Resources."
4. United Nations Centre for Human Settlements et al., "Water and Sustainable Urban Development and Drinking Water Supply and Sanitation in the Urban Context," background paper prepared for the International Conference on Water and the Environment: Development Issues for the 21st Century, Dublin, Ireland, January 26–31, 1992.
5. 1.2 billion figure from Joseph Christmas and Carel de Rooy, "The Decade and Beyond: At a Glance," *Water International*, September 1991; 80 percent figure from G.A. Brown, "Keynote Address," in *World Water 1986*, proceedings (London: Thomas Telford Ltd., 1987).

6. Zbigniew Bochniarz, "Water Management Problems in Economies in Transition," *Natural Resources Forum*, February 1992.
7. Maharashtra example from World Bank, *India: Irrigation Sector Review, Vol. I* (Washington, D.C.: 1991).

CHAPTER 2.   Signs of Scarcity

1. One cubic kilometer equals 1 billion cubic meters or 1 trillion liters; in standard U.S. usage, the equivalent is about 264 billion gallons or 810,190 acre-feet. R.L. Nace, U.S. Geological Survey, 1967, as cited in Frits van der Leeden et al., *The Water Encyclopedia* (Chelsea, Mich.: Lewis Publishers, Inc., 1990).
2. These are rounded estimates based on M.I. L'vovitch, *World Water Resources and Their Future* (Washington, D.C.: American Geophysical Union, 1979), as cited in van der Leeden et al., *The Water Encyclopedia*.
3. Global runoff is assessed at 40,673 cubic kilometers by the former Institute of Geography, National Academy of Sciences of the Soviet Union, as published in World Resources Institute (WRI), *World Resources 1992–93* (New York: Oxford University Press, 1992); population from Population Reference Bureau (PRB), *1992 World Population Data Sheet* (Washington, D.C.: 1992); stable supply from L'vovitch, *World Water Resources and Their Future*.
4. U.N. Department of International Economic and Social Affairs, *World Population Prospects 1990* (New York: 1991); WRI, *World Resources 1992–93*.
5. Swedish hydrologist Malin Falkenmark has put forth this definition, which has gradually gained wider use. See, for example, Malin Falkenmark, "The Massive Water Scarcity Now Threatening Africa—Why Isn't it Being Addressed?" *Ambio*, Vol. 18, No. 2, 1989.
6. PRB, *1992 World Population Data Sheet*; WRI, *World Resources 1992–93*; the four African countries that will be added to the water-scarce category are Malawi, Morocco, South Africa, and Sudan.
7. Population projections from PRB, *1992 World Population Data Sheet*.
8. Abdulla Ali Al-Ibrahim, "Excessive Use of Groundwater Resources in Saudi Arabia: Impacts and Policy Options," *Ambio*, Vol. 20, No. 1, 1991.
9. Mark Nicholson, "Subsidised Security," *Financial Times*, January 30, 1992; Mark Nicholson, "Saudis Reap Bumper Wheat Subsidy," *Financial Times*, January 21, 1992; International Mon-

etary Fund, *International Financial Statistics* (Washington, D.C.: April 1992).

10. Al-Ibrahim, "Excessive Use of Groundwater Resources in Saudi Arabia"; "Green Revolution in the Desert," *Pakistan and Gulf Economist*, August 20–26, 1988; "Survey: The Arab World," *The Economist*, May 12, 1990.

11. "Gadaffi Turns His Pipedream into Reality," *Financial Times*, August 29, 1991; Fred Pearce, "Will Gaddafi's Great River Run Dry?" *New Scientist*, September 7, 1991; Hugh Roberts, "Deep Waters Run Still," *South*, August 1991; Peter F.M. McLoughlin, "Libya's Great Manmade River Project: Prospects and Problems," *Natural Resources Forum*, August 1991.

12. Pearce, "Will Gaddafi's Great River Run Dry?"

13. Depletion estimates from Pearce, "Will Gaddafi's Great River Run Dry?" and from Roberts, "Deep Waters Run Still."

14. John B. Weeks, "High Plains Regional Aquifer-System Study," in Ren Jen Sun, ed., *Regional Aquifer-System Analysis Program of the U.S. Geological Survey: Summary of Projects, 1978–84* (Washington, D.C.: U.S. Government Printing Office, 1986).

15. Depletion figure from tables supplied by the High Plains Underground Water Conservation District No. 1, Lubbock, Tex., dated May 3, 1991; Texas water use, preliminary estimates, from Wayne Solley, Water Use Information, U.S. Geological Survey, Reston, Va., private communication, April 27, 1992, with final estimates to appear in *Estimated Water Use of the United States in 1990* (Washington, D.C.: U.S. Government Printing Office, forthcoming); irrigated area figures from Texas Water Development Board (TWDB), *Surveys of Irrigation in Texas—1958, 1964, 1969, 1974, 1979, 1984, and 1989* (Austin, Tex.: 1991), and from Comer Tuck, TWDB, Austin, Tex., private communication, November 26, 1991.

16. PRB, *1992 Population Data Sheet*; WRI, *World Resources 1992–93*; James E. Nickum, "Beijing's Rural Water Use," prepared for East-West Center North China Project, Honolulu, Hawaii, March 1987; The Chinese Research Team for Water Resources Policy and Management in Beijing-Tianjin Region of China, *Report on Water Resources Policy and Management for the Beijing-Tianjin Region of China* (Beijing: Sino-US Cooperative Research Project on Water Resources Policy and Management, 1987); "Water Rules Tightened; Fines Levied," *China Daily*, May 18, 1989; North China Plain grain output from Frederick W. Crook, *Agricultural Statistics of the People's Republic of China, 1949–86* (Washington, D.C.: Economic Research Service, U.S. Department of Agriculture, 1988); Li Hong, "Beijing Set to

Tackle Water Thirst," *China Daily*, October 17, 1989; "North-ern, Coastal Area Cities Face Water Shortages," *China Daily*, August 29, 1991, as reprinted in *JPRS Report: Environmental Issues*, October 11, 1991.

17. People's Republic of China, State Science and Technology Commission, *Beijing-Tianjin Water Resources Study: Final Report* (Beijing: 1991).

18. 80 percent figure from M.A. Chitale, "Comprehensive Manage-ment of Water Resources: India's Achievements and Perspec-tives," prepared for World Bank International Workshop on Comprehensive Water Resources Management Policies, Wash-ington, D.C., June 24–28, 1991; Jayanto Bandyopadhyay, "The Ecology of Drought and Water Scarcity," *The Ecologist*, Vol. 18, No. 2, 1988; Jayanto Bandyopadhyay, "Riskful Confusion of Drought and Man-Induced Water Scarcity," *Ambio*, Vol. 18, No. 5, 1989.

19. M.G. Chandrakanth and Jeff Romm, "Groundwater Depletion in India—Institutional Management Regimes," *Natural Re-sources Journal*, Summer 1990.

20. Ibid.

CHAPTER 3.   Engineering's Promise

1. K.H.S. Gunatilaka and L.U. Weerakoon, "Evolution of Water Management in Sri Lanka," in International Water Resources Association (IWRA), *Water for World Development: Proceedings of the VIth IWRA World Congress on Water Resources*, Vol. III (Ur-bana, Ill.: 1988).

2. "Large" dams are those over 15 meters high. Number of dams and rate of dam construction from World Resources Institute (WRI), *World Resources 1992–93* (New York: Oxford University Press, 1992), and from data from the International Commission on Large Dams and the *International Water Power and Dam Con-struction Handbook* as presented in Frits van der Leeden et al., *The Water Encyclopedia* (Chelsea, Mich: Lewis Publishers, Inc., 1990); Nagara project from Steven R. Weisman, "As One More Dam is Built, Japanese Anger Bursts," *New York Times*, March 14, 1991; *Japan Environment Monitor*, April 30, 1990; Michael Cross, "Japanese River Scheme Survives Barrage of Criticism," *New Scientist*, April 11, 1992.

3. World water use from I.A. Shiklomanov, "Global Water Re-sources," *Nature & Resources*, Vol. 26, No. 3, 1990; Mississippi River flow from van der Leeden et al., *The Water Encyclopedia*; WRI, *World Resources 1992–93*; Population Reference Bureau,

1992 World Population Data Sheet (Washington, D.C.: 1992); historical population data from U.N. Department of International Economic and Social Affairs, *World Population Prospects 1990* (New York: 1991).

4. Van der Leeden et al., *The Water Encyclopedia*; WRI, *World Resources 1992–93*; U.S. reservoir storage from Robert M. Hirsch, Assistant Chief Hydrologist, Research and External Coordination, U.S. Geological Survey, presentation before National Women's Democratic Club, Washington, D.C., October 31, 1991; U.S. Congressional Budget Office, *Efficient Investments in Water Resources: Issues and Options* (Washington, D.C.: U.S. Government Printing Office, 1983). See also Gilbert F. White, "Water Resource Adequacy: Illusion and Reality," in Julian Simon and Herman Kahn, eds., *The Resourceful Earth* (New York: Basil Blackwell, Inc., 1984).

5. Daniel F. Luecke, "Controversy over Two Forks Dam," *Environment*, May 1990; Philip Shabecoff, "E.P.A. Office Recommends Against Dam Project Near Denver," *New York Times*, March 27, 1990; Michael Weisskopf, "EPA's Reilly to Veto Dam: Effects of Denver Project 'Unacceptable'," *Washington Post*, November 23, 1990; "Denver Suburbs Ponder Lawsuit," *U.S. Water News*, October 1991; filing of lawsuit from Daniel F. Luecke, Senior Scientist, Environmental Defense Fund, Boulder, Colo., July 1, 1992.

6. Philip P. Micklin, "Soviet River Diversion Projects: Problems and Prospects," in IWRA, *Water for World Development*, Vol. I.

7. Ibid.; V.M. Kotlyakov, "The Aral Sea Basin: A Critical Environmental Zone," *Environment*, January/February 1991; "Uzbek, Tajik Presidents Propose Return to Siberian River Diversion," *Moscow News*, May 12–19, 1991, as reprinted in *JPRS Report: Environmental Issues*, September 13, 1991.

8. Frank Quinn, "Interbasin Water Diversions: A Canadian Perspective," *Journal of Soil and Water Conservation*, November/December 1987; D.J. Gamble, "Is the GRAND Canal Scheme in Canada's Interest?" in IWRA, *Water for World Development*, Vol. I; Frank Quinn, "Large-Scale Water Transfers," in ibid.

9. "NAWAPA Is Still Talked About Despite Obvious Obstacles," *U.S. Water News*, January 1989; Gamble, "Is the GRAND Canal Scheme in Canada's Interest?"; Quinn, "Large-Scale Water Transfers"; Sandra Postel, "U.S. Should Refuse Canadian Water Offer," *Journal of Commerce*, June 10, 1985.

10. U.S. Office of Technology Assessment (OTA), *Alaskan Water for California? The Subsea Pipeline Option—Background Paper* (Washington, D.C.: U.S. Government Printing Office, 1992).

11. Zhang Zezhen and Chen Zhikai, "Drought and Water Shortage in Northern China and Their Countermeasures," in IWRA, *Water for World Development*, Vol. IV; "River Diversion Project Viewed," *China Daily*, August 10, 1989.

12. Facts and quote from "Water Crisis Looms in China," *World Water and Environmental Engineer*, March 1992; inability of the diversion to meet year-2000 needs is from People's Republic of China, State Science and Technology Commission, *Beijing-Tianjin Water Resources Study: Final Report* (Beijing: 1991).

13. *Public Papers of the Presidents of the United States, John F. Kennedy, 1961* (Washington, D.C.: U.S. Government Printing Office, 1962).

14. Klaus Wangnick Consulting, *1990 IDA Worldwide Desalting Plants Inventory* (Englewood, N.J.: International Desalination Association, 1990); Shiklomanov, "Global Water Resources."

15. OTA, *Using Desalination Technologies for Water Treatment—Background Paper* (Washington, D.C.: U.S. Government Printing Office, March 1988); average urban water charge, 25¢ per cubic meter, from World Water/World Health Organization, *The International Drinking Water Supply and Sanitation Decade Directory* (London: Thomas Telford Ltd., 1987); U.S. Department of Agriculture (USDA), Economic Research Service (ERS), *Economic Indicators of the Farm Sector, Cost of Production—Major Field Crops, 1989* (Washington, D.C.: April 1991); Bill McBride, USDA, ERS, private communication, August 8, 1991.

16. Wangnick Consulting, *Worldwide Desalting Plants*; "Achievements of Water Desalination Program Lauded," *Al-Riyad*, June 27, 1991, as reprinted in *JPRS Report: Environmental Issues*, September 13, 1991.

17. California Coastal Commission, *Seawater Desalination in California*, preliminary draft report, March 29, 1991; Harriet Miller, City Council Member, Santa Barbara, Calif., testimony before the Senate Environment and Public Works Committee, July 23, 1991; Robert Reinhold, "Hit-or-Miss Rainfall Brings Relief to Part of California," *New York Times*, April 26, 1992.

18. OTA, *Using Desalination*; Wangnick Consulting, *Worldwide Desalting Plants*; "Congress Hears Merits of Desalination," *U.S. Water News*, September 1991.

CHAPTER 4.   Bread and Water

1. U.N. Food and Agriculture Organization (FAO), *1990 Production Yearbook* (Rome: 1991), adjusted for the United States and Tai-

wan with irrigated area data from, respectively, U.S. Department
of Agriculture (USDA), Economic Research Service (ERS), *Agricultural Resources, Cropland, Water and Conservation*, September
1991, and Sophia Hung, USDA, ERS, private communication,
June 21, 1991.

2. Estimated irrigated area in 1900 was 48 million hectares, according to K.K. Framji and I.K. Mahajan, *Irrigation and Drainage in
the World: A Global Review* (New Delhi, India: Caxton Press Private Limited, 1969); irrigated area today from FAO, *Production
Yearbook*, and from USDA; population from Population Reference Bureau (PRB), *World Population Estimates and Projections by
Single Years: 1750–2100* (Washington, D.C.: 1992); harvest estimate from W. Robert Rangeley, "Irrigation and Drainage in the
World," in Wayne R. Jordan, ed., *Water and Water Policy in World
Food Supplies* (College Station, Tex.: Texas A&M University
Press, 1987).

3. Figure 4–1 based on FAO, *Production Yearbook*, adjusted for the
United States and Taiwan with data from USDA, ERS, and on
United Nations, Department of International Economic and Social Affairs, *World Population Prospects, 1990* (New York: 1991);
FAO, *Production Yearbook*.

4. Per capita grain trend from USDA, ERS, *World Grain Database*
(unpublished printouts) (Washington, D.C.: 1991), and from
Francis Urban and Michael Trueblood, *World Population by
Country and Region, 1950–2050* (Washington, D.C.: USDA,
ERS, 1990).

5. India figures from Mark Svendsen, "Sources of Future Growth in
Indian Irrigated Agriculture," presented at the Planning Workshop on Policy Related Issues in Indian Irrigation, Ootacamund,
Tamil Nadu, India, April 26–28, 1988; China estimate from
Daniel Gunaratnum, China Agriculture Operations Division,
World Bank, Washington, D.C., private communication, June
20, 1989; supporting figures and Mexico estimate from Robert
Repetto, *Skimming the Water: Rent-Seeking and the Performance of
Public Irrigation Systems*, WRI Paper 4 (Washington, D.C.: World
Resources Institute, 1986); Brazil figure from Jean-Louis Ginnsz,
Brazil Agriculture Operations Division, World Bank, private
communication, June 7, 1989; Rangeley, "Irrigation and Drainage in the World"; Thayer Scudder, "Conservation Vs. Development: River Basin Projects in Africa," *Environment*, March 1989;
FAO, *Consultation on Irrigation in Africa* (Rome: 1987). See also
Montague Yudelman, "Sustainable and Equitable Development
in Irrigated Environments," in H. Jeffrey Leonard et al., *Environment and the Poor: Development Strategies for a Common Agenda*

(New Brunswick, N.J.: Transaction Books for Overseas Development Council, 1989).

6. Funding trends from G. Levine et al., "Irrigation in Asia and the Near East in the 1990s: Problems and Prospects," prepared for the Irrigation Support Project for Asia and the Near East at the request of the Asia/Near East Bureau, U.S. Agency for International Development, Washington, D.C., August 1988.

7. 150 million figure from M.E. Jensen et al., "Irrigation Trends in World Agriculture," in B.A. Stewart and D.R. Nielsen, eds., *Irrigation of Agricultural Crops* (Madison, Wisc.: American Society of Agronomy, 1990).

8. V.A. Kovda, "Loss of Productive Land Due to Salinization," *Ambio*, Vol. 12, No. 2, 1983.

9. W. Robert Rangeley, Berkshire, U.K., private communication, January 30, 1989; reference to World Bank study from Shawki Barghouti and Guy Le Moigne, "Irrigation and the Environmental Challenge," *Finance & Development*, June 1991; Mexico figure from Yudelman, "Sustainable and Equitable Development in Irrigated Environments."

10. James Rhoades, U.S. Salinity Laboratory, Riverside, Calif., private communication, September 1, 1989; Soviet figure from Philip P. Micklin, Western Michigan Univeristy, Kalamazoo, Mich., private communication, October 13, 1989; Barghouti and Le Moigne, "Irrigation and the Environmental Challenge"; rate of irrigation expansion from Sandra Postel, *Water for Agriculture: Facing the Limits*, Worldwatch Paper 93 (Washington, D.C.: Worldwatch Institute, December 1989).

11. Clifford Dickason, "Improved Estimates of Groundwater Mining Acreage," *Journal of Soil and Water Conservation*, May/June 1988; Clifford Dickason, USDA, ERS, Washington, D.C., private communication, October 19, 1989; "Advance Census Reports Show Irrigation Rebound," *Agricultural Outlook*, May 1989.

12. James E. Nickum and John Dixon, "Environmental Problems and Economic Modernization," in Charles E. Morrison and Robert F. Dernberger, *Focus: China in the Reform Era*, Asia-Pacific Report 1989 (Honolulu, Hawaii: East-West Center, 1989); reference to Tamil Nadu in Carl Widstrand, ed., *Water Conflicts and Research Priorities* (Elmsford, N.Y.: Pergamon Press, 1980); Raj Chengappa, "India's Water Crisis," *India Today*, May 31, 1986, excerpted in *World Press Review*, August 1986.

13. Svendsen, "Sources of Future Growth in Indian Irrigated Agriculture"; irrigation potential from Shri C.G. Desai, "Planning

Targets for Irrigation Development," presented at the Planning Workshop on Policy Related Issues in Indian Irrigation, Ootacamund, Tamil Nadu, India, April 26–28, 1988; Gandhi quoted in Omar Sattaur, "India's Troubled Waters," *New Scientist*, May 27, 1989.

14. Number of dams from Sattaur, "India's Troubled Waters"; J. Patel, "Who Benefits Most from Damming the Narmada," *Economic and Political Weekly* (India), December 29, 1990; Omar Sattaur, "Fair Deal Denied to People Displaced by Dam," *New Scientist*, August 3, 1991.

15. Patel, "Who Benefits Most from Damming the Narmada"; Baba Amte, "What Price the Big Dams?" in The Hindu, *Survey of the Environment, 1991* (Madras, India: M/s. Kasturi & Sons Ltd., undated); 1.8 million figure from M.A. Chitale, "Comprehensive Management of Water Resources: India's Achievements and Perspectives," prepared for World Bank International Workshop on Comprehensive Water Resources Management Policies, Washington, D.C., June 24–28, 1991; Baba Amte, *Cry, the Beloved Narmada* (Chandrapur, Maharashtra, India: Maharogi Sewa Samiti, 1989).

16. Barbara Crossette, "Water, Water Everywhere? Many Now Say 'No!' " *New York Times*, October 7, 1989; "World Bank to Assess Narmada," *World Rivers Review*, March/April 1991.

17. Bradford Morse and Thomas R. Berger, *Sardar Sarovar*, Report of the Independent Review (Ottawa, Ont.: Resource Futures International, Inc., 1992).

18. Barghouti and Le Moigne, "Irrigation and the Environmental Challenge."

19. Jose Olivares, "The Potential for Irrigation Development in Sub-Saharan Africa," in Shawki Barghouti and Guy Le Moigne, *Irrigation in Sub-Saharan Africa: The Development of Public and Private Systems* (Washington, D.C.: World Bank, 1990).

20. PRB, *World Population Estimates and Projections by Single Years: 1750–2100*; 1 billion figure from World Bank, *World Development Report 1991* (New York: Oxford University Press, 1991).

21. Brian Forster, "Wheat Can Take On More Than a Pinch of Salt," *New Scientist*, December 3, 1988; Israeli use of salty water from Uri Or, Kibbutz Magal, Israel, private communication, March 2, 1992. For information on salt-loving crops, see U.S. National Research Council, *Saline Agriculture: Salt-Tolerant Plants for Developing Countries* (Washington, D.C.: National Academy Press, 1990).

22. World Bank, *India: Irrigation Sector Review*, Vol. I (Washington, D.C.: 1991).

CHAPTER 5.   Paradise Lost

1. Alexei Yablokov, Member of Parliament and Deputy Chairman, Committee of Ecology, Supreme Soviet, private communication in Washington, D.C., June 18, 1991.
2. Philip P. Micklin, *The Water Management Crisis in Soviet Central Asia*, The Carl Beck Papers in Russian and East European Studies (Pittsburgh, Pa.: University of Pittsburgh, 1991).
3. V.M. Kotlyakov, "The Aral Sea Basin: A Critical Environmental Zone," *Environment*, January/February 1991; Micklin, *The Water Management Crisis in Soviet Central Asia*.
4. Typhoid and hepatitis rates from Kotlyakov, "The Aral Sea Basin"; other adverse health effects from Micklin, *The Water Management Crisis in Soviet Central Asia*.
5. "Uzbek, Tajik Presidents Propose Return to Siberian River Diversion," *Moscow News*, May 12–19, 1991, as reprinted in *JPRS Report: Environmental Issues*, September 13, 1991; Yusup S. Kamalov, Deputy Director of the Union in Defense of the Aral and Amu Dar'ya, private communication in Washington, D.C., October 1991.
6. Damien Lewis, "Will Botswana Put Diamonds Before the Environment?" *New African*, July 1991; Neil Henry, "Arid Botswana Keeps Its Democracy Afloat," *Washington Post*, March 21, 1991.
7. Henry, "Arid Botswana Keeps Its Democracy Afloat."
8. "Okavango Delta Threatened by Boro River Diversion," *World Rivers Review*, March/April 1991, as reprinted in *Ecoafrica*, June 1991; Lewis, "Will Botswana Put Diamonds Before the Environment?"; Gwenda Brophy, Botswana country profile in *New Internationalist*, October 1991; David B. Ottaway, "A Second Look Saves a Great Delta," *Washington Post*, June 18, 1992.
9. Shawki Barghouti and Guy Le Moigne, *Irrigation in Sub-Saharan Africa: The Development of Public and Private Systems* (Washington, D.C.: World Bank, 1990).
10. E.A.A. Zaki, "Water Resource Management: Sudan," prepared for World Bank International Workshop on Comprehensive Water Resources Management Policies, Washington, D.C., June 24–28, 1991; M.A. Abu-Zeid and M.A. Rady, "Egypt's Water Resources Management and Policies," prepared for World Bank International Workshop; size of swamp from Barghouti and Le Moigne, *Irrigation in Sub-Saharan Africa*; opposition to the project from Dale Whittington and Elizabeth McClelland, "Opportunities for Regional and International Co-

operation in the Nile Basin," University of North Carolina at Chapel Hill, June 1991.

11. Glossy ibis population during the early dry season from Whittington and McClelland, "Opportunities for Regional and International Cooperation in the Nile Basin."

12. Whittington and McClelland, "Opportunities for Regional and International Cooperation in the Nile Basin."

13. Robert Arnsberger, assistant superintendent, Everglades National Park, Fla., private communication, November 2, 1990.

14. Area of wetlands remaining from John Lancaster, "Monumental Salvage Job Is Planned for the Everglades," *Washington Post*, February 20, 1990; drop in wading-bird population from Arnsberger, private communication, and from Nicole Duplaix, "South Florida Water: Paying the Price," *National Geographic*, July 1990; endangered species from K. Michael Fraser, "Huge Environment Project Aims to Save Florida's Everglades," *Christian Science Monitor*, July 30, 1991.

15. Agricultural land acreage and effects of contaminated runoff from Duplaix, "South Florida Water." See also Jeffrey Schmalz, "Pollution Poses Growing Threat to Everglades," *New York Times*, September 17, 1989.

16. For a discussion of the 1988 lawsuit and July 1991 settlement, see Duplaix, "South Florida Water," and Fraser, "Huge Environment Project Aims to Save Florida's Everglades"; Everglades Protection Act taxing district from Ann Overton, "In the Swim: A Life Preserver for Florida's Threatened Water Bodies," *Waterlines* (South Florida Water Management District), Summer 1991; expansion of eastern Everglades and role of Army Corp of Engineers from Michael Satchell, "Can the Everglades Still Be Saved?" *U.S. News and World Report*, April 2, 1990.

17. Fraser, "Huge Environment Project Aims to Save Florida's Everglades"; *Everglades Connection* (South Florida Water Management District), May 1992; Florida residents' water usage from Duplaix, "South Florida Water."

18. California's 1990 population, 39.8 million, from U.S. Bureau of the Census, *Statistical Abstract of the United States 1991* (Washington, D.C.: 1991); Canada's 1990 population, 26.6 million, from Population Reference Bureau, *1990 World Population Data Sheet* (Washington, D.C.: 1990); $2.1 billion figure from John Lancaster, "Drought Adds Urgency To California Water Debate," *Washington Post*, July 13, 1991.

19. Decline in salmon numbers from Marc Reisner, "Can Anyone Win This Water War?" *National Wildlife*, June/July 1991; delta

smelt from Jane Gross, "A Dying Fish May Force California to Break Its Water Habits," *New York Times*, October 27, 1991; Tom Kenworthy, "Plan to Protect Smelt Could Threaten California Water System," *Washington Post*, September 28, 1991; Charles McCoy, "U.S. to Propose Listing Rare Smelt As Threatened," *Wall Street Journal*, September 27, 1991; Charles McCoy, "Lobbyists' Smelt-and-Bird Campaign Is Assault Against Endangered Species Act Itself, Some Say," *Wall Street Journal*, August 29, 1991.

20. Tom Harris, *Death in the Marsh* (Covelo, Calif.: Island Press, 1991).

21. Damage due to toxics in agricultural drainage water from Tom Harris, "A Valley Filled with Selenium," *Sacramento Bee*, July 16, 1989; The Wilderness Society, "Ten Most Endangered National Wildlife Refuges," Washington, D.C., October 1988; Eliot Marshall, "High Selenium Levels Confirmed in Six States," *Science*, January 10, 1986; Harris, *Death in the Marsh*.

22. Harris, *Death in the Marsh*.

23. Jack E. Williams et al., "Fishes of North America Endangered, Threatened, or of Special Concern: 1989," *Fisheries*, November/December 1989; continental fish, crayfish, and mussel figures from Larry Master, "Aquatic Animals: Endangerment Alert," *Nature Conservancy*, March/April 1991.

24. Salvador Contreras Balderas, "Conservation of Mexican Freshwater Fishes: Some Protected Sites and Species, and Recent Federal Legislation," in W.L. Minckley and James E. Deacon, eds., *Battle Against Extinction: Native Fish Management in the American West* (Tucson, Ariz.: The University of Arizona Press, 1991).

25. Ibid.

26. Willa Nehlsen et al., "Pacific Salmon at the Crossroads: Stocks at Risk from California, Oregon, Idaho, and Washington," *Fisheries*, March/April 1991; John Davies, "Columbia River Barges Spared in Latest Fish Rescue Proposal," *Journal of Commerce*, November 26, 1991; Rocky Barker, "U.S. Fish Agency Takes the Slow Road," *High Country News*, July 1, 1991; expected listing of Snake River chinook as endangered from Jay M. Sheppard, Division of Endangered Species, U.S. Fish and Wildlife Service, Washington, D.C., private communication, July 2, 1992.

27. Timothy Egan, "Fight to Save Salmon Starts Fight Over Water," *New York Times*, April 1, 1991; Timothy Egan, "U.S. Proposes Listing a Salmon as Endangered," *New York Times*, April 3, 1991; Charles McCoy, "Salmon Battle Could Spawn

Much Bitterness," *Wall Street Journal,* June 5, 1991; Davies, "Columbia River Barges Spared."
28. Holmes quoted in Joseph L. Sax, "The Constitution, Property Rights and the Future of Water Law," Western Water Policy Project Discussion Series Paper No. 2, Natural Resources Law Center, University of California, Berkeley, 1990.

CHAPTER 6. Hydropolitics

1. Quoted in "Water Scarcity, Quality in Africa Aggravated by Augmented Population Growth," *International Environmental Reporter,* October 1989.
2. 40 percent figure from Evan Vlachos, "Water, Peace and Conflict Management," *Water International,* Vol. 15, No. 4, 1990; African basins from Asit K. Biswas, "Water for Sustainable Development in the 21st Century: A Global Perspective," Address to the VIIth World Congress on Water Resources, Rabat, Morocco, May 13, 1991.
3. Figure 6–1 from *The Times Atlas of the World,* 7th ed. (New York: Times Books, 1985); Thomas Naff, "The Jordan Basin: Political, Economic, and Institutional Issues," prepared for World Bank International Workshop on Comprehensive Water Resources Management Policies, Washington, D.C., June 24–28, 1991; Maher F. Abu Taleb et al., "Water Resources Planning and Development in Jordan: Problems, Future Scenarios, Recommendations" (draft), World Bank, Washington, D.C., June 1991; population estimates from Population Reference Bureau (PRB), *1992 World Population Data Sheet* (Washington, D.C.: 1992); King Hussein reference from Joyce R. Starr, "Nature's Own Agenda: A War for Water in the Mideast," *Washington Post,* March 3, 1991.
4. 25–40 percent figure from Joyce R. Starr, "Water Wars," *Foreign Policy,* March 1991; Naff, "The Jordan Basin"; Fred Pearce, "Wells of Conflict on the West Bank," *New Scientist,* June 1, 1991.
5. Pearce, "Wells of Conflict on the West Bank"; Starr, "Nature's Own Agenda."
6. 20 percent figure from Daniel Zaslavsky, Israeli Water Commissioner, Tel Aviv, private communication, March 5, 1992; "Pollution, Salinity Affecting Domestic Water Sources," *Jerusalem Post,* June 20, 1991, as reprinted in *JPRS Report: Environmental Issues,* July 9, 1991; Israel Ministry of the Environment, "State Comptroller Report: The Water Quantity Crisis," *Israel Environment Bulletin,* Spring 1991; Naff, "The Jordan Basin."

7. Population from PRB, *1992 World Population Data Sheet*; source of Egypt's water from Raj Krishna, "The Legal Regime of the Nile River Basin," in Joyce R. Starr and Daniel C. Stoll, eds., *The Politics of Scarcity: Water in the Middle East* (Boulder, Colo: Westview Press, 1988).

8. M.A. Abu-Zeid and M.A. Rady, "Egypt's Water Resources Management and Policies," presented at World Bank International Workshop on Comprehensive Water Resources Management Policies, Washington, D.C., June 24–28, 1991.

9. Jonglei project from Dale Whittington and Elizabeth McClelland, "Opportunities for Regional and International Cooperation in the Nile Basin," University of North Carolina at Chapel Hill, June 1991.

10. Reduced flow figure from U.S. Agency for International Development-Cairo, Office of Irrigation and Land Development, "Irrigation Briefing Paper," April 12, 1987; Sadat quote from Starr, "Water Wars."

11. Scot E. Smith and Hussam M. Al-Rawahy, "The Blue Nile: Potential for Conflict and Alternatives for Meeting Future Demands," *Water International*, Vol. 15, No. 4, 1990; Egypt blocking project from Alan Cowell, "Now, a Little Steam. Later, Maybe a Water War," *New York Times*, February 7, 1990.

12. Peter Rogers, Harvard University, "International River Basins: Pervasive Undirectional Externalities," presented at conference on The Economics of Transnational Commons, Universita di Siena, Italy, April 25-27, 1991.

13. Whittington and McClelland, "Opportunities for Regional and International Cooperation in the Nile Basin"; Ethiopia's irrigated area in 1989 was 162,000 hectares, according to U.N. Food and Agriculture Organization, *1990 Production Yearbook* (Rome: 1991).

14. Sahim Tekeli, "Turkey Seeks Reconciliation for the Water Issue Induced by the Southeastern Anatolia Project (GAP)," *Water International*, Vol. 15, No. 4, 1990.

15. "Send for the Dowsers," *The Economist*, December 16, 1989; John Kolars, "The Future of the Euphrates River," prepared for World Bank International Workshop on Comprehensive Water Resources Management Policies, Washington, D.C., June 24–28, 1991; Syrian population growth from PRB, *1992 World Population Data Sheet*; Starr, "Water Wars."

16. Ozal quote from Thomas Goltz, "Turkey Diverts Euphrates River," *Washington Post*, January 14, 1990; Ozal's threat from

ibid. and from Clyde Haberman, "Dam is Watering Hope for a New Fertile Crescent," *New York Times,* March 30, 1990; Starr, "Water Wars."

17. Presentation by His Excellency Necati Utkan, Turkish Ambassador to Iraq, at a press conference on the Middle East Water Summit held by the Global Water Summit Initiative, Washington, D.C., March 19, 1991; Starr, "Water Wars"; Kolars, "The Future of the Euphrates River."

18. Starr, "Water Wars"; Tekeli, "Turkey Seeks Reconciliation for the Water Issue Induced by the Southeastern Anatolia Project (GAP)"; cost estimate from "Bridging the GAP," *World Water and Environmental Engineer,* April 1992; Trilateral Commission from Starr, "Water Wars."

19. Nahid Islam, "The Ganges Water Dispute: Environmental and Related Impacts on Bangladesh," *BIISS Journal,* Vol. 12, No. 3, 1991.

20. Ibid.; Sheila Tefft, "India and Bangladesh at Odds Over Water as Rivers Run Low," *Christian Science Monitor,* June 7, 1988; Rahman Jahangir, "Indo-Bangla Water Talks Begin," *Green File* (Centre for Science and Environment, New Delhi), January 1989.

21. Islam, "The Ganges Water Dispute"; Tefft, "India and Bangladesh at Odds Over Water as Rivers Run Low."

22. Stephen McCaffrey, "International Organizations and the Holistic Approach to Water Problems," *Natural Resources Journal,* Winter 1991; "Draft Articles on the Law of Non-navigational Uses of International Watercourses," in Preparatory Committee for the U.N. Conference on Environment and Development, "Development of Legal Instruments for Transboundary Waters," progress report by the Secretariat prepared for the Third Session, Geneva, August 12-September 4, 1991.

23. M. Yunus Khan, "Boundary Water Conflict Between India and Pakistan," *Water International,* Vol. 15, No. 4, 1990; Jagat S. Mehta, "The Indus Water Treaty: A Case Study in the Resolution of an International River Basin Conflict," *Natural Resources Forum,* Vol. 12, No. 1, 1988.

24. Syed S. Kirmani, "Water, Peace and Conflict Management: The Experience of the Indus and Mekong River Basins," *Water International,* Vol. 15, No. 4, 1990; Mehta, "The Indus Water Treaty."

25. Senegal example from Shawki Barghouti and Guy Le Moigne, *Irrigation in Sub-Saharan Africa: The Development of Public and*

Private Systems (Washington, D.C.: World Bank, 1990); U.N. Environment Programme, "An Action Plan for the Zambezi," UNEP Environment Brief No. 6, Nairobi, Kenya, undated.

CHAPTER 7.   A World Heating Up

1. Paul E. Waggoner, ed., *Climate Change and U.S. Water Resources* (New York: John Wiley & Sons, 1990). For general background, see Stephen H. Schneider, *Global Warming: Are We Entering the Greenhouse Century?* (San Francisco, Calif.: Sierra Club Books, 1989).
2. Intergovernmental Panel on Climate Change, *Policymakers' Summary of the Potential Impacts of Climate Change: Report from Working Group II to IPCC* (Geneva: World Meteorological Organization/U.N. Environment Programme, 1990); Paul E. Waggoner, "U.S. Water Resources Versus an Announced But Uncertain Climate Change," *Science*, March 1, 1991.
3. Peter H. Gleick and Linda Nash, *The Societal and Environmental Costs of the Continuing California Drought* (Oakland, Calif.: Pacific Institute for Studies in Development, Environment, and Security, 1991).
4. Ibid.
5. Ibid.
6. Ibid.
7. John C. Schaake, "From Climate to Flow," in Waggoner, *Climate Change and U.S. Water Resources*. For a similar analysis, see Peter H. Gleick, "Regional Hydrologic Consequences of Increases in Atmospheric CO_2 and Other Trace Gases," *Climatic Change*, Vol. 10, 1987.
8. P.H. Gleick, "Observed Changes in Regional Variability of Runoff in the Western United States," Eighth Annual Pacific Climate Workshop on Climate Variability of the Eastern North Pacific and Western North America, Asilomar, Calif., March 10–13, 1991; original runoff data from Maurice Roos, "Possible Climate Change and Its Impact on Water Supply in California," presented at the Oceans '89 Conference, Seattle, Wash., September 20, 1989.
9. James E. Hansen, NASA Goddard Institute for Space Studies, "Modeling Greenhouse Climate Effects," testimony before Subcommittee on Science, Technology, and Space, Committee on Commerce, Science, and Transportation, U.S. Senate, Washington, D.C., May 8, 1989.
10. Norman J. Rosenberg et al., "From Climate and CO_2 Enrichment to Evapotranspiration," in Waggoner, *Climate Change and*

U.S. Water Resources; Richard M. Adams, "Global Climate Change and U.S. Agriculture," *Nature*, May 17, 1990. See also Fred Pearce, "High and Dry in the Global Greenhouse," *New Scientist*, November 10, 1990.

11. Rosenberg et al., "From Climate and CO_2 Enrichment to Evapotranspiration"; Adams, "Global Climate Change and U.S. Agriculture"; Fakhri A. Bazzaz and Eric D. Fajer, "Plant Life in a CO_2-Rich World," *Scientific American*, January 1992.

12. Pearce, "High and Dry in the Global Greenhouse."

13. Assumes range of $2,000–4,000 per hectare; costs in some areas could be substantially higher. Sandra Postel, *Water for Agriculture: Facing the Limits*, Worldwatch Paper 93 (Washington, D.C.: Worldwatch Institute, December 1989); $1 billion figure from Operations and Evaluation Division, World Bank, Washington, D.C., private communication, July 1, 1992.

14. Harry E. Schwarz and Lee A. Dillard, "Urban Water," in Waggoner, *Climate Change and U.S. Water Resources*.

15. Peter P. Rogers and Myron B. Fiering, "From Flow to Storage," in Waggoner, *Climate Change and U.S. Water Resources*.

CHAPTER 8.   Thrifty Irrigation

1. Water use estimates from I.A. Shiklomanov, "Global Water Resources," *Nature & Resources*, Vol. 26, No. 3, 1990.

2. Efficiency estimate from W.R. Rangeley, "Irrigation and Drainage in the World," presented at the International Conference on Food and Water, College Station, Tex., May 26–30, 1985. See also Mohamed T. El-Ashry et al., "Salinity Pollution from Irrigated Agriculture," *Journal of Soil and Water Conservation*, January/February 1985.

3. For a good overview of irrigation efficiency, see E.G. Kruse and D.F. Heermann, "Implications of Irrigation System Efficiencies," *Journal of Soil and Water Conservation*, November/December 1977; see also Marvin E. Jensen, "Irrigation Research and Development in the Next Decade," reprinted from IDRC-90 Proceedings, Lethbridge, Alba., Canada, July 1990.

4. "District Salutes Water Savings By Area Irrigators," *The Cross Section* (High Plains Underground Water Conservation District No. 1, Lubbock, Tex.), November 1989; payback from Ken Carver, High Plains Underground Water Conservation District No. 1, private communication, May 29, 1992. For other surge results, see Richard Bartholomay, "USDI Funds Study: Surge Irrigation Lowers Salt Loading in Colorado River," *Irrigation Journal*, September/October 1991.

5. Donald H. Negri and John J. Hanchar, *Water Conservation Through Irrigation Technology* (Washington, D.C.: Economic Research Service (ERS), U.S. Department of Agriculture (USDA), November 1989); Carver, private communication, March 24, 1992.

6. For more on LEPA, see William M. Lyle and James P. Bordovsky,"LEPA: Low Energy Precision Application," *Irrigation Journal*, April 1991; retrofit costs and payback periods from Carver, private communication, May 24, 1992.

7. "Irrigation System Upgrade Provides Producers with Substantial Water, Fuel Savings," *The Cross Section* (High Plains Underground Water Conservation District No. 1, Lubbock, Tex.), December 1990.

8. Depletion chart of the Ogallala aquifer from Wayne Wyatt, Manager, High Plains Underground Water Conservation District No. 1, "Water Management—Southern High Plains of Texas," unpublished, May 1991; "District Salutes Water Savings by Area Irrigators"; Texas Water Development Board, *Surveys of Irrigation in Texas—1958, 1964, 1969, 1974, 1979, 1984, and 1989* (Austin, Tex.: 1991); the actual time periods for comparing the depletion rate were 1966–71 and 1986–91.

9. Quote from Meir Ben-Meir, Director General, Ministry of Agriculture of Israel, "Irrigation—Establishing Research Priorities," address, April 1988. See also "Israel's Water Policy: A National Commitment," in U.S. Congress, Office of Technology Assessment, *Water-Related Technologies for Sustainable Agriculture in Arid/Semi-Arid Lands: Selected Foreign Experience* (Washington, D.C.: U.S. Government Printing Office, 1983).

10. 1974 estimate from Don Gustafson, "Drip Irrigation in the World—State of the Art," in *Israqua '78: Proceedings of the International Conference on Water Systems and Applications* (Tel Aviv: Israel Centre of Waterworks Appliances, 1978); Table 8–1 from Dale Bucks, Microirrigation Working Group, International Commission on Irrigation and Drainage (ICID), Beltsville, Md., private communication, June 22, 1992, with irrigated area from U.N. Food and Agriculture Organization (FAO), *1990 Production Yearbook* (Rome: 1991), with adjustments from USDA for the United States and Taiwan.

11. J.S. Abbott, "Micro Irrigation—World Wide Usage," *ICID Bulletin*, January 1984; costs from David Melamed, "Technological Developments in Irrigation: The Israeli Experience," unpublished paper, and from Paul Wilson et al., *Drip Irrigation for Cotton: Implications for Farm Profits* (Washington, D.C.: USDA, 1984); 130,000 figure from Bucks, private communication. For

general background and basic features of drip irrigation, see Kobe Shoji, "Drip Irrigation," *Scientific American*, November 1977, and Sterling Davis and Dale Bucks, "Drip Irrigation," in Claude H. Pair et al., eds., *Irrigation* (Silver Spring, Md.: The Irrigation Association, 1983).

12. Areas and crops under drip irrigation from Bucks, private communication; world irrigated area from FAO, *1990 Production Yearbook*, with adjustments from USDA, ERS.

13. Israel's drip area from Bucks, private communication; Israel's total irrigated area from FAO, *1990 Production Yearbook*; water efficiency gain from Jehoshua Schwarz, "Israel Water Sector Review: Past Achievements, Current Problems and Future Options," prepared for the World Bank by Tahal Consulting Engineers Ltd, Tel Aviv, Israel, December 1990; topping out of water use efficiency from Jiftah Ben-Asher, Institute for Desert Research, Ben Gurion University, Be'er Sheva, Israel, presented at the International Seminar on Efficient Water Use, Mexico City, October 1991.

14. Dan Rymon and Uri Or, "Advanced Technologies in Traditional Agriculture (ATTA): A New Approach," *ICID Bulletin*, Vol. 39, No. 1, 1990; author's visit to the Jiftlik Valley, March 3, 1992.

15. Rymon and Or, "Advanced Technologies in Traditional Agriculture"; author's visit to the region and Uri Or, private communication, March 3, 1992.

16. Author's visit to the region and Or, private communication; Gaza and Negev experiments mentioned by Saul Arlosoroff, Project Manager, Water and Sanitation, World Bank, Washington, D.C., private communication, May 27, 1992.

17. Robert Chambers, *Managing Canal Irrigation: Practical Analysis from South Asia* (Cambridge: Cambridge University Press, 1988).

18. Ibid.

19. Sri Lanka example from Rathnasiri Ekanayake et al., *A Rapid-Assessment Survey of the Irrigation Component of the Anuradhapura Dry-Zone Agriculture Project (ADZAP)* (Colombo, Sri Lanka: International Irrigation Management Institute, 1990); Mexico example from Ronald Cummings et al., *Waterworks: Improving Irrigation Management in Mexican Agriculture*, WRI Paper No. 5 (Washington, D.C.: World Resources Institute, December 1989).

20. Montague Keen, "Clearer Thoughts Flow on Irrigation," *Ceres*, May/June 1988. See also Romana P. de los Reyes and Sylvia Ma. G. Jopillo, *An Evaluation of the Philippine Participatory Com-*

munal Irrigation Program (Quezon City: Institute of Philippine Culture, Ateneo de Manila University, 1986).

21. Shaul Manor et al., eds., *Role of Social Organizers in Assisting Farmer-Managed Irrigation Systems*, Proceedings of a regional workshop of the Farmer-Managed Irrigation Systems Network, Khon Kaen, Thailand, May 15–20, 1989 (Colombo, Sri Lanka: International Irrigation Management Institute, 1990); International Irrigation Management Institute, *Managing Irrigation in the 1990's: A Brief Guide to the Strategy of the International Irrigation Management Institute* (Colombo, Sri Lanka: 1989).

22. U. Gautam, "Role of Social Organizers in Improving Irrigation Management: The Experience in Nepal," in Manor et al., *Role of Social Organizers in Assisting Farmer-Managed Irrigation Systems*.

23. Comisión Nacional del Agua, *Water Policies and Strategies* (Mexico City: December 1990); see also Enrique Palacios Velez, "Irrigation Systems in Mexico," in *Irrigation in Latin America: Present Situation, Problem Areas and Areas of Potential Improvement* (Colombo, Sri Lanka: International Irrigation Management Institute, 1990); total irrigated area from FAO, *Production Yearbook 1990*; 2 million figure from Hector Garduno, Instituto Mexicano de Tecnología del Agua, Comisión Nacional del Agua, private communication, Mexico City, October 20, 1991.

24. Peter Rogers et al., *Eastern Waters Study: Strategies to Manage Flood & Drought in the Ganges-Brahmaputra Basin* (Arlington, Va.: Irrigation Support Project for Asia and the Near East, 1989). See also D.J.W. Berkoff, *Irrigation Management on the Indo-Gangetic Plain* (Washington, D.C.: World Bank, 1990).

25. Ken Carver et al., "Irrigating by the Block," *Water Management Note*, High Plains Underground Water Conservation District No. 1, Lubbock, Tex., undated; for an overview of soil moisture monitoring methods, see Tad Weems, "Survey of Moisture Measurement Instruments," *Irrigation Journal*, January/February 1991; test results from Gail Richardson, *Saving Water from the Ground Up* (New York: INFORM, Inc., 1985).

26. California Department of Water Resources, *Water Conservation News*, October 1991, and various other issues; Holly Sheradin, CIMIS Program Manager, Water Conservation Office, California Department of Water Resources, Sacramento, Calif., private communication, March 31, 1992; Greenhill Farms example from Robert D. Hof and Eric Schine, "Drought is the Mother of Invention," *Business Week*, October 14, 1991; Gloria Pacheco, CIMIS, Sacramento, Calif., private communication, May 27, 1992.

CHAPTER 9. Small-Scale Solutions

1. Discussion with visiting delegation from several Sahelian countries, including Sekou Haidara, Advisor to the General Director of Water Works and Energy, Republic of Mali; Jorge Cabral, Director-Geral dos Assuntos Politicos Economicos e Culturais, Ministerio dos Negocios Estrangeiros, Guinea-Bissau; Tamsier D. M'Bye, Under Secretary, Ministry of External Affairs of the Gambia; and Pape Samba Mboup, Attache de Cabinet du Ministre de l'Integration Economique Africaine, Senegal; Washington, D.C., September 13, 1991.
2. Percentage of cropland irrigated from U.N. Food and Agriculture Orgnaization (FAO), *1990 Production Yearbook* (Rome: 1991), with adjustments from U.S. Department of Agriculture, Economic Research Service; figures on arid and semiarid lands and their inhabitants from H.M. Lovenstein et al., "Runoff Agroforestry in Arid Lands," *Forest Ecology and Management*, Vol. 45, 1991.
3. Shawki Barghouti and Guy Le Moigne, *Irrigation in Sub-Saharan Africa: The Development of Public and Private Systems* (Washington, D.C.: World Bank, 1990).
4. Crop failures one in three years from Barghouti and Le Moigne, *Irrigation in Sub-Saharan Africa*.
5. For an overview of some traditional methods and their use, see Chris Reij, *Indigenous Soil and Water Conservation in Africa* (London: International Institute for Environment and Development, 1991); Will Critchley, *Looking After Our Land: Soil and Water Conservation in Dryland Africa* (Oxford: Oxfam, 1991).
6. Critchley, *Looking After Our Land.*
7. World Bank, *Vetiver Grass (Vetiveria zizanioides): A Method of Vegetative Soil and Moisture Conservation* (New Delhi: 1987).
8. Peru terracing from Christiaan Gischler and C. Fernandez Jauregui, "Low-Cost Techniques for Water Conservation and Management in Latin America," *Nature and Resources*, July/September 1984.
9. Critchley, *Looking After Our Land.*
10. Ibid.
11. National Academy of Sciences, *More Water for Arid Lands: Promising Technologies and Research Opportunities* (Washington, D.C.: 1974); U.N. Environment Programme, *Rain and Stormwater Harvesting in Rural Areas* (Dublin: Tycooly International Publishing Ltd., 1983).
12. Author's visit to Avdat and discussions with Pedro Berliner,

Head, Runoff Agriculture Unit, Jacob Blaustein Institute for Desert Research, Ben Gurion University of the Negev, Sde Boqer Campus, Israel, March 4, 1992.

13. Berliner, private communication; Lovenstein et al., "Runoff Agroforestry in Arid Lands."

14. Mark Svendsen and Ruth Meinzen-Dick, "Garden Irrigation: The Invisible Sector," International Food Policy Research Institute (IFPRI), Washington, D.C., unpublished paper, October 1990.

15. A.M. Michael (former Director, Indian Agricultural Research Institute), "Raising Yield in Rainfed Lands: Stress on Water Management," in The Hindu, Survey of Indian Agriculture 1990 (Madras, India: M/s. Kasturi & Sons Ltd., undated). See also K. Palanisami, "Tank Irrigation In South India: What Next?" Irrigation Management Network (Overseas Development Institute, London), July 1990.

16. 37 percent figure from Mark Svendsen et al., "Choice of Irrigation Technology in Zimbabwe," in Structural Change in African Agriculture, IFPRI Policy Briefs 5 (Washington, D.C.: 1990).

17. Peter H. Stern, Small-Scale Irrigation: A Manual of Low-cost Water Technology (London: Intermediate Technology Publications Ltd., 1979); Svendsen and Meinzen-Dick, "Garden Irrigation."

18. Svendsen and Meinzen-Dick, "Garden Irrigation"; Zimbabwe's irrigated area from FAO, 1990 Production Yearbook; 2.5-hectare plot example from Ian Scoones, "Wetlands in Drylands: Key Resources for Agricultural and Pastoral Production in Africa," Ambio, December 1991.

19. Svendsen and Meinzen-Dick, "Garden Irrigation."

20. Ibid.

21. Ellen P. Brown and Robert Nooter, "Successful Small-scale Irrigation in the Sahel" (draft), World Bank, Washington, D.C., September 1991; Richard Carter, ed., NGO Casebook on Small Scale Irrigation in Africa (Rome: FAO, 1989); Barghouti and Le Moigne, Irrigation in Sub-Saharan Africa.

22. Barghouti and Le Moigne, Irrigation in Sub-Saharan Africa.

23. Brown and Nooter, "Successful Small-scale Irrigation in the Sahel."

24. Carter, NGO Casebook on Small Scale Irrigation in Africa.

25. Brown and Nooter, "Successful Small-scale Irrigation in the Sahel."

CHAPTER 10. Wastewater No More

1. Author's visit to western Galilee region and discussion of project with Danny Sherban, Yodfat Consulting Engineers, Yodfat, Israel, March 1, 1992.
2. Example from John R. Sheaffer and Leonard A. Stevens, *Future Water: An Exciting Solution to America's Most Serious Resource Crisis* (New York: William Morrow & Company, Inc., 1983).
3. Hillel I. Shuval et al., *Wastewater Irrigation in Developing Countries: Health Effects and Technical Solutions* (Washington, D.C.: World Bank, 1986).
4. Ibid.
5. C.R. Bartone and S. Arlosoroff, "Irrigation Reuse of Pond Effluents in Developing Countries," *Water Science Technology*, Vol. 19, No. 12, 1987.
6. Shaul Streit, Project Director, Israel Sewerage Project, Tel Aviv, Israel, private communication, March 5, 1992; Jehoshua Schwarz, "Israel Water Sector Review: Past Achievements, Current Problems and Future Options," prepared for the World Bank by Tahal Consulting Engineers Ltd., Tel Aviv, Israel, December 1990.
7. Streit, private communication; Shaul Streit, "On-Land Treatment and Disposal of Municipal Sewage Agro-Sanitary Integration: The Israeli Experience," prepared for World Bank Seminar, Washington, D.C., March 19, 1992.
8. Bartone and Arlosoroff, "Irrigation Reuse of Pond Effluents in Developing Countries"; Hillel I. Shuval, *Wastewater Irrigation in Developing Countries: Health Effects and Technical Solutions* (Washington, D.C.: World Bank, 1990).
9. Shuval, *Wastewater Irrigation in Developing Countries*.
10. Calculation is from ibid., and assumes average water use of 175 liters per person per day, which is low for most industrial countries but a reasonable figure for moderate- to low-income countries. It also assumes that 85 percent of the water used is released to the sewer system, which yields the 150-liter-per-day figure, and that the average irrigation application rate is 10,000 cubic meters per hectare a year, a figure that obviously varies with the type and efficiency of irrigation system.
11. Hillel I. Shuval, "The Development of Water Reuse in Israel," *Israel Environment Bulletin*, Summer 1991; Streit, private communication.
12. Herman Bouwer, "Agricultural and Municipal Use of Wastewater," prepared for meeting of the International Association of

Water Pollution Research and Control, Washington, D.C., May 1992; Shuval, *Wastewater Irrigation in Developing Countries*. See also Asit K. Biswas and Abdullah Arar, eds., *Treatment and Reuse of Wastewater* (London: Butterworths, 1988).

13. Shuval et al., *Wastewater Irrigation in Developing Countries*; Bartone and Arlosoroff, "Irrigation Reuse of Pond Effluents in Developing Countries." A project to treat the sewage flows is now under construction, according to Saul Arlosoroff, Project Manager, Water and Sanitation, World Bank, Washington, D.C., private communication, May 27, 1992.

14. Henk W. de Koning, ed., *Environmental Health and the Management of Fresh Water Resources in the Americas*, prepared for Pan American Health Organization and World Health Organization, Washington, D.C., January 1992; Shuval et al., *Wastewater Irrigation in Developing Countries*; author's visit to Tula Irrigation District, Mexico, October 1991.

15. Bartone and Arlosoroff, "Irrigation Reuse of Pond Effluents in Developing Countries."

16. David Richard et al., "Wastewater Reclamation Costs and Water Reuse Revenue," prepared for American Water Resources Association 1991 Summer Symposium, Water Supply and Water Reuse: 1991 and Beyond, San Diego, Calif., June 6–10, 1991.

17. Cost of primary and secondary treatment assumed to be $99 per acre-foot, from Bouwer, "Agricultural and Municipal Use of Wastewater"; incremental costs of advanced waste treatment from Richard et al., "Wastewater Reclamation Costs and Water Reuse Revenue."

18. Bahman Sheikh and Don Marske, "Planning for Water Reuse to Meet the Growth Needs of the 21st Century for the City of Los Angeles," prepared for American Water Resources Association 1991 Summer Symposium, Water Supply and Water Reuse: 1991 and Beyond, San Diego, Calif., June 6–10, 1991; James A. Van Haun and Martin G. Rigby, "Water Reclamation: A Key Component of Orange County Water District's Groundwater Management Plan," in *Proceedings of CONSERV 90: The National Conference and Exposition Offering Water Supply Solutions for the 1990s* (Columbus, Ohio: National Ground Water Association, 1990); Kirke Guild, "Wastewater Reuse in Tucson, Arizona—Past, Present, and Future," in ibid.

19. Kenneth T. Holmes et al.,"City of Phoenix, 23rd Avenue/RID Water Reuse Project," in *Proceedings of CONSERV 90*.

20. Joseph V. Towry and David Shulmister, "Water Conservation Pioneers," *Quality Cities '90*, May 1990; Utility Accounts Divi-

sion, City of St. Petersburg, "Quick Facts about Services and Rates" (rates effective October 1, 1990).

21. D.E. Bourne and G.S. Watermeyer, "Proposed Potable Reuse—An Epidemiological Study in Cape Town," in *Proceedings of the Water Reuse Symposium II* (Denver, Colo.: AWWA Research Foundation, 1981); Bouwer, "Agricultural and Municipal Use of Wastewater"; William C. Lauer, "More Efficient Use of Limited Water Resources: Direct Potable Reuse—A Denver Perspective," *U.S. Water News*, November 1989.

22. Costs from Bouwer, "Agricultural and Municipal Use of Wastewater," and from Lauer, "More Efficient Use of Limited Water Resources: A Denver Perspective."

CHAPTER 11. Industrial Recycling

1. Roy Opie, "Prevention Is Not Always Better Than Cure," *World Water and Environmental Engineer*, October 1990; Gary Allie, American Iron and Steel Institute, Washington, D.C., private communication, April 27, 1992.

2. I.A. Shiklomanov, "Global Water Resources," *Nature & Resources*, Vol. 26, No. 3, 1990; World Resources Institute, *World Resources 1992–93* (New York: Oxford University Press, 1992). Hydropower is not included in these industrial use figures, since it does not require removing water from a river or lake, but it does compete with the protection of fisheries, aquatic habitat, and recreational values.

3. Wayne Solley et al., "Preliminary Water Use Estimates in the United States During 1990," U.S. Geological Survey, Open File Report 92–63, Washington, D.C., July 1992; Allie, private communication.

4. Figure 11–1 from National Land Agency, Water Resources Department, *Water Resources in Japan: Present State of Water Resources Development, Conservation and Utilization* (Tokyo: various years), from International Monetary Fund (IMF), *1991 Yearbook* (Washington, D.C.: 1991), from IMF, *International Financial Statistics* (Washington, D.C.: June 1992), and from World Bank, *World Develoment Report 1992* (New York: Oxford University Press, 1992).

5. 1950 figure from Wayne Solley et al., *Estimated Use of Water in the United States in 1985* (Washington, D.C.: U.S. Government Printing Office, 1988); Solley et al., "Preliminary Water Use Estimates"; IMF, *International Financial Statistics* (Washington, D.C.: various years); projection for 2000 from Culp/Wesner/ Culp, *Water Reuse and Recycling: Evaluation of Needs and Potential*,

Vol. 1 (Washington, D.C.: U.S Department of the Interior, 1979). The last census is documented in U.S. Department of Commerce, Bureau of the Census, *Water Use in Manufacturing* (Washington, D.C.: U.S. Government Printing Office, 1981).

6. Roy Opie, "Germany's Double Bill," *World Water and Environmental Engineer*, April 1991; IMF, *International Financial Statistics* (various years); Opie, "Prevention Is Not Always Better Than Cure"; Sandra Postel, *Conserving Water: The Untapped Alternative*, Worldwatch Paper 67 (Washington, D.C.: Worldwatch Institute, September 1985).

7. William Sweetman, Spalding Sports Worldwide, Chicopee, Mass., private communication, March 18, 1992; Massachusetts Water Resources Authority, "MWRA Honors Spalding Sports Worldwide for Water Conservation," news release, Boston, Mass., September 27, 1991.

8. "Bylaw 1–90, Just the Beginning of Savings," *Industrial Perspectives*, January 1992; *Environews*, Fall 1991 (both publications of the Regional Municipality of Waterloo).

9. Mark Manzione et al., "California Industries Cut Water Use," *Journal of the AWWA*, October 1991; Table 11–1 from City of San Jose, Brown and Caldwell Consultants, and Department of Water Resources, *Case Studies of Industrial Water Conservation in the San Jose Area* (Sacramento: California Department of Water Resources, 1990); calculation of households served assumes 1 acre-foot (1,234 cubic meters) meets the annual needs of two average households. For more examples, see Maggie Murphy, "Industrial Water Conservation is Feasible," *Water Conservation News* (California Department of Water Resources), April 1991.

10. William W. Wade et al., *Cost of Industrial Water Shortages*, prepared by Spectrum Economics, Inc., for the California Urban Water Agencies (San Francisco, Calif.: 1991); California's economic ranking from U.S. Central Intelligence Agency, *Handbook of Economic Statistics, 1991* (Washington, D.C.: 1991).

11. Wade et al., *Cost of Industrial Water Shortages*.

12. Ibid.

13. Ibid.

14. Egyptian Program for Water Resources Management Sector, "National Projects for Environmental Protection and Development" (draft), Terms of Reference, Third Project, Cairo, November 1991; David E. Sanger, "Chemical Leak in Korea Brings Forth a New Era," *New York Times*, April 16, 1991; see also "Doosan Company Contaminates Source of Tap Water Again," (Seoul) *YONHAP*, April 23, 1991, as reprinted in

JPRS Report: Environmental Issues, May 31, 1991; "Severe Contamination of Rimac River Reported," (Madrid) *EFE*, June 23, 1991, as reprinted in *JPRS Report: Environmental Issues*, July 9, 1991.

15. Choo Wai Chan, Public Utilities Board-Singapore, private communication, Singapore, May 23, 1992.

16. Ramesh Bhatia and Malin Falkenmark, "Water Resource Policies and the Urban Poor: Innovative Approaches and Policy Imperatives," prepared for the International Conference on Water and the Environment: Development Issues for the 21st Century, Dublin, Ireland, January 26–31, 1992.

CHAPTER 12. Conserving in Cities

1. Author's visit to Mexico City, October 1991.

2. Table 12–1 is from Worldwatch Institute, based on the following: for Mexico City, Boston, and Waterloo, see notes to paragraphs on these cities later in this chapter; Jerusalem from A.D. Rosenberg, Deputy Manager, Department of Water Supply and Sewerage, Municipality of Jerusalem, private written communication, Jerusalem, Israel, May 18, 1992; Southern California from Matthew Puffer, Metropolitan Water District, Los Angeles, Calif., private communication, June 8, 1992; Beijing from "Beijing Water Shortages Prompt Introduction of Regulations," *China Daily*, October 30, 1991, as reprinted in *JPRS Report: Environmental Issues*, January 13, 1992; Singapore from World Health Organization, Regional Working Group on Water Supply Management, *Country Report: Singapore* (Kuala Lumpur, Malaysia: 1990); Bogor from Ramesh Bhatia and Malin Falkenmark, "Water Resource Policies and the Urban Poor: Innovative Approaches and Policy Imperatives," prepared for the International Conference on Water and the Environment: Development Issues for the 21st Century, Dublin, Ireland, January 26–31, 1992; Melbourne from Melbourne Water Resources Review, "Water for Our Future," Issues Paper, June 1991.

3. *Agua 2000: Estrategia para la Ciudad de México*, May 1991; overpumping rate from Juan Manuel Martinez Garcia, Director General of Hydraulic Construction and Operation, Mexico City, private communication, October 21, 1991; population size is the rough midpoint of 15 million estimate in the latest official Mexican census and the 20.2 million estimate by the U.N. Department of International Economic and Social Affairs, *World Urbanization Prospects 1990* (New York: 1991), according to Kimberly Crews, Director of Population Education Programs, Population

Reference Bureau, Washington, D.C., private communication, May 14, 1992.

4. *Agua 2000*; Departamento del Distrito Federal, Secretaría General de Obras, Dirección General de Construcción y Operación Hidráulica, *Memoria: Programa de Uso Eficiente del Agua* (Mexico City: 1991); Martinez Garcia, private communication; Mexico City's population growth rate is 3 percent per year, according to official Mexican census estimate, as related by Crews, private communication.

5. Juan Manuel Martinez Garcia, *Program de Uso Eficiente del Agua en la Ciudad de México* (Mexico City: Ciudad de México DDF, 1991); Departamento del Distrito Federal, *Reglamento del Servicio de Agua y Drenaje para el Distrito Federal* (Mexico City: 1990); Martinez Garcia, private communication.

6. Martinez Garcia, *Programa de Uso Eficiente del Agua en la Ciudad de México*; Martinez Garcia, private communication.

7. Number of residents is author's calculation based on water use of 300 liters per day, from Martinez Garcia, private communication.

8. Jim Robertson, University of Waterloo, Waterloo, Ont., Canada, private communication, March 25, 1992; Ralph Luhowy, Regional Municipality of Waterloo, Waterloo, Ont., Canada, private communication, May 1, 1992.

9. Luhowy, private communication.

10. "Regulations to Amend Ontario Regulation 815/84 Made Under the Ontario Water Resources Act," *The Ontario Gazette*, O. Reg. 134/92; Ministry of Natural Resources, "Ontario Announces Strategy To Reduce Water Consumption and Use Water Wisely," News Release, Toronto, Ont., Canada, August 19, 1991.

11. Barbara Jordan, "Door-to-Door Water Conservation Retrofits: The San Jose Story," *Home Energy*, July/August 1990.

12. City goal and estimated flow reduction from City of San Jose, "Water Conservation Program Analysis, Evaluation and Long-Term Planning Study," final report, prepared by James M. Montgomery Consulting Engineers Inc., Walnut Creek, Calif., October 1991; Jordan, "Door-to Door Retrofits"; Barbara Jordan, Jordan and Company, private communication, Concord, Calif., May 6, 1992.

13. For a good overview of trends in the United States, see Amy Vickers, "The Emerging Demand-Side Era in Water Management," *Journal of the AWWA*, October 1991; Paul F. Levy and

William A. Brutsch, *MWRA Long Range Water Supply Program* (Boston: Massachusetts Water Resources Authority (MWRA), 1990).

14. "Boston Delays Water Diversion Plan," *New York Times*, November 16, 1986.

15. Levy and Brutsch, *MWRA Long Range Water Supply Program*; Paul Dinatale, MWRA, Boston, Mass., private communication, June 4, 1992; Marcis Kempe, MWRA, Boston, Mass., private communication, June 4, 1992.

16. Figure 12–1 from Kempe, private communication, June 5–6, 1992; cost comparison from Paul Levy, Executive Director, MWRA, presentation at a briefing sponsored by Energy and Environmental Study Institute, Washington, D.C., December 16, 1991; quote from Levy and Brutsch, *MWRA Long Range Water Supply Program*.

17. Price example from Ramesh Bhatia and Malin Falkenmark, "Water Resource Policies and the Urban Poor: Innovative Approaches and Policy Imperatives," prepared for the International Conference on Water and the Environment: Development Issues for the 21st Century, Dublin, Ireland, January 26–31, 1992.

18. Canadian figures from David B. Brooks et al., "Pricing: A Neglected Tool for Managing Water Demand," *Alternatives*, Vol. 17, No. 3, 1990; British example from "Water, Water—at a Price," *The Economist*, April 13, 1991. For a general discussion of pricing, see Roger McNeill and Donald Tate, *Guidelines for Municipal Water Pricing*, Social Science Series No. 25 (Ottawa, Canada: 1991).

19. Edmonton-Calgary example from Brooks et al., "Pricing: A Neglected Tool for Managing Water Demand"; U.K. figure from "Water, Water—at a Price."

20. William E. Martin et al., *Saving Water in a Desert City* (Washington, D.C.: Resources for the Future, 1984); $75 million figure from Linda Smith, "Tucson: A Water Ethic in the Desert," *U.S. Water News*, July 1990.

21. Bhatia and Falkenmark, "Water Resource Policies and the Urban Poor."

22. "Mass. Mandates Low-Flow," *U.S. Water News*, March 1989; states that have adopted the 6-liter toilet standard from Amy Vickers, Amy Vickers & Associates, Boston, Mass., private communication, June 16, 1992. The 15 states are California, Connecticut, Delaware, Georgia, Maryland, Massachusetts,

Nevada, New Jersey, New York, North Carolina, Oregon, Rhode Island, Texas, Utah, and Washington. Colorado has adopted standards for showerheads and faucets, but not the 6-liter toilet standard.

23. Amy Vickers, "Water-Use Efficiency Standards for Plumbing Fixtures: Benefits of National Legislation," *Journal of the AWWA*, May 1990; legislative action from Vickers, private communication, May 22, 1992.

24. Conserv 90, "Xeriscape: A Growing Idea in Water Conservation," news release, Dublin, Ohio, February 27, 1990; Patricia Wellingham-Jones, "The Dry Garden Comes of Age," *Garden*, July/August 1986; Novato example from John Olaf Nelson, "Water Conserving Landscapes Show Impressive Savings," in *Proceedings of CONSERV 90: The National Conference and Exposition Offering Water Supply Options for the 1990s* (Columbus, Ohio: National Ground Water Association, 1990).

25. Raymond Uecker, Executive Director, National Xeriscape Council, Roswell, Ga., private communication, April 1, 1992; Tucson ordinance from Dan Charles, "Squeezing the Deserts Dry," *New Scientist*, September 14, 1991.

26. Lagos figure from Peter Rogers, "Integrated Urban Water Resources Management," keynote paper, International Conference on Water and the Environment: Development Issues for the 21st Century, Dublin, Ireland, January 26–31, 1992; other cities from Bhatia and Falkenmark, "Water Resource Policies and the Urban Poor."

27. MWRA, *MWRA at Work: Massachusetts Water Resources Authority Annual Report 1990* (Boston, Mass.: 1991); "Leak Detectives Boost Manila Supply," *World Water*, November 1983; Jakarta example from Bhatia and Falkenmark, "Water Resource Policies and the Urban Poor."

28. Population Reference Bureau (PRB), *World Population Estimates and Projections by Single Years: 1750–2100* (Washington, D.C.: 1992); PRB, *1992 World Population Data Sheet* (Washington, D.C.: 1992).

29. Cost estimate from Saul Arlosoroff, Project Manager, Water and Sanitation, World Bank, Washington, D.C., private communication, May 27, 1992.

30. World Bank programs from Arlosoroff, private communication.

CHAPTER 13.   Pricing, Markets, and Regulations

1. Benjamin Franklin, *Poor Richard's Almanac*, as quoted in John Bartlett, *Bartlett's Familiar Quotations*, 14th ed. (Boston, Mass.: Little, Brown and Company, 1968).
2. Ronald Cummings et al., *Waterworks: Improving Irrigation Management in Mexican Agriculture*, WRI Paper 5 (Washington, D.C.: World Resources Institute, December 1989); World Bank, *Indonesia: Sustainable Development of Forests, Land, and Water* (Washington, D.C.: World Bank, 1990); Pakistan example from Robert Repetto, *Skimming the Water: Rent-Seeking and the Performance of Public Irrigation Systems*, WRI Paper 4 (Washington, D.C.: World Resources Institute, December 1986); Asit K. Biswas, "Land and Water Management for Sustainable Agricultural Development in Egypt: Opportunities and Constraints," Government of Egypt, Ministry of Agriculture and Land Reclamation, and U.N. Food and Agriculture Organization, February 1991.
3. World Bank, *India: Irrigation Sector Review*, Vol. I (Washington, D.C.: 1991); M.A. Chitale, "Comprehensive Management of Water Resources: India's Achievements and Perspectives," prepared for World Bank International Workshop on Comprehensive Water Resources Management Policies, Washington, D.C., June 24–28, 1991.
4. U.S. Department of the Interior, Bureau of Reclamation, *1987 Summary Statistics Vol. 1, Water, Land, and Related Data* (Denver, Colo.: 1988); Richard W. Wahl, *Markets for Federal Water: Subsidies, Property Rights, and the Bureau of Reclamation* (Washington, D.C.: Resources for the Future, 1989).
5. Wahl, *Markets for Federal Water*.
6. Michael R. Moore and Catherine A. McGuckin, "Program Crop Production and Federal Irrigation Water," in U.S. Department of Agriculture (USDA), Economic Research Service, *Agricultural Resources: Cropland, Water, and Conservation Situation and Outlook Report*, Washington, D.C., September 1988.
7. Islamic norms from Biswas, "Land and Water Management for Sustainable Agricultural Development in Egypt."
8. Montague Keen, "Clearer Thoughts Flow on Irrigation," *Ceres*, May/June 1988.
9. John Ambler, Program Officer, Ford Foundation, New Delhi, India, private communication in Washington, D.C., April 2,

1992. See also S.N. Lele and R.K. Patel, "Working for farmer participation in Irrigation Management in Major Irrigation Projects: A Report on Pilot Study in Minor 7 of Mula Project (1987–1991)," Centre for Applied Systems Analysis in Development in collaboration with Irrigation Department, Government of Maharashtra, Pune, India, November 1991.

10. India energy subsidies from World Bank, *India: Irrigation Sector Review.* See also Marcus Moench, *Drawing Down the Buffer: Upcoming Ground Water Issues in India* (Oakland, Calif.: Pacific Institute for Studies in Development, Environment, and Security, 1991).

11. Sandra Postel, "California's Liquid Deficit," *New York Times*, February 27, 1991. See also Marc Reisner and Sarah Bates, *Overtapped Oasis: Reform or Revolution for Western Water* (Washington, D.C.: Island Press, 1990).

12. Robert Reinhold, "Farmers in West May Sell Something More Valuable Than Any Crop: Water," *New York Times*, April 6, 1992; "The Environmental Senate," *Washington Post*, April 3, 1992.

13. Zhang Zezhen et al., "Challenges to and Opportunities for Development of China's Water Resources in the 21st Century," *Water International*, March 1992; Jehoshua Schwarz, "Israel Water Sector Review: Past Achievements, Current Problems and Future Options," prepared for World Bank by Tahal Consulting Engineers Ltd, Tel Aviv, Israel, December 1990; India example from World Bank, *India: Irrigation Sector Review*, and from Ambler, private communication.

14. Rodney T. Smith and Roger Vaughan, eds., "1991 Annual Transaction Review: Water Comes to Town," *Water Strategist* (Stratecon, Inc., Claremont, Calif.), January 1992.

15. 7 percent figure from Deborah Moore and Zach Willey, "Water in the American West: Institutional Evolution and Environmental Restoration in the 21st Century," *Colorado Law Review*, Vol. 62, No. 4, 1991.

16. "Conservation and Drought Strategies," *Water Market Update*, December 1988.

17. Elizabeth Checchio, *Water Farming: The Promise and Problems of Water Transfers in Arizona* (Tucson: University of Arizona, 1988); Gary Thacker, USDA extension agent, College of Agriculture, University of Arizona, Tucson, private communication, July 29, 1991; "Arizona Rewrites Groundwater Law," *Water Strategist* (Stratecon, Inc., Claremont, Calif.), July 1991.

18. Bangladesh example from World Bank, "Water Resources Management—A Policy Paper" (draft), Washington, D.C.,

May 11, 1992; Tushaar Shah, "Water Markets and Irrigation Development in India," *Indian Journal of Agricultural Economics*, July/September 1991.

19. Shah, "Water Markets and Irrigation Development in India."
20. "Arizona Rewrites Groundwater Law," *Water Strategist* (Stratecon, Inc., Claremont, Calif.), July 1991.
21. Ibid.
22. Smith and Vaughan, "1991 Annual Transaction Review."
23. Wildlife refuge example from "Central Valley Bill Up in Bradley Panel," *Weekly Bulletin* (Energy and Environmental Study Institute, Washington, D.C.,) May 6, 1991.
24. Moore and Willey, "Water in the American West"; Matthew J. McKinney et al., "The Protection of Instream Flows in Montana: A Legal-Institutional Perspective," in Lawrence J. MacDonnell et al., eds., *Instream Flow Protection in the West* (Boulder, Colo.: University of Colorado School of Law, 1989).
25. Ellen Sullivan Casey, "Water Law—Public Trust Doctrine," *Natural Resources Journal*, July 1984; Harrison C. Dunning, "A New Front in the Water Wars: Introducing the 'Public Trust' Factor," *California Journal*, May 1983; Robert Buderi, "New Plan for Mono Lake," *Nature*, October 12, 1989; "Los Angeles Loses Again in Mono Lake Case," *U.S. Water News*, February 1990; "CA: Court Continues Injunction Against Diversions from Mono Lake," *Water Intelligence Monthly*, May 1991; Finney quoted in "Judge Rules for Mono, Awards Fees," *The Mono Lake Newsletter* (The Mono Lake Committee, Lee Vining, Calif.), Fall 1991.
26. Maynard M. Hufschmidt and David S. McCauley, "Water Resources Management in a River/Lake Basin Context: A Conceptual Framework with Examples from Developing Countries," *Water Resources Development*, December 1988; $6 billion figure from K. Mahmood, *Reservoir Sedimentation: Impact, Extent, and Mitigation* (Washington, D.C.: World Bank, 1987).
27. John B. Doolette and William B. Magrath, eds., *Watershed Development in Asia: Strategies and Technologies* (Washington, D.C.: World Bank, 1990).
28. Ibid.
29. "New York's Suburbs Urged to Plan for Their Groundwater Future," *The Groundwater Newsletter* (Water Information Center, Inc., Plainview, N.Y.), February 28, 1991; Sarah Meyland, Executive Director, Citizens Campaign for the Environment, Massapequa, N.Y., private communication, April 20, 1992. See also Sarah J. Meyland, "Watershed Management Advances Using State-of-the-Art Technologies and Strategies," in *Pro-*

ceedings of CONSERV 90: The National Conference and Exposition Offering Water Supply Solutions for the 1990s (Columbus, Ohio: National Ground Water Association, 1990).

30. Leslie Sauer, "Making a Habit of Restoration: Saving the Eastern Deciduous Forest," Andropogon Associates, Ltd., Philadelphia, Pa., unpublished, October 1990; Long Island Regional Planning Board, *Nonpoint Source Management Handbook* (Hauppauge, N.Y.: 1984).

31. Richard W. Robbins et al., "Effective Watershed Management for Surface Water Supplies," *Journal of the AWWA*, December 1991.

CHAPTER 14.   A Water Ethic

1. Several of the concepts and ideas in this chapter echo those of Aldo Leopold in his essay "The Land Ethic," in Aldo Leopold, *A Sand County Almanac* (New York: Oxford University Press, 1949).

2. M.A. Abu-Zeid and M.A. Rady, "Egypt's Water Resources Management and Policies," prepared for World Bank International Workshop on Comprehensive Water Resources Management Policies, Washington, D.C., June 24–28, 1991; Population Reference Bureau (PRB), *1992 World Population Data Sheet* (Washington, D.C.: 1992).

3. Cost estimate from Joseph Christmas and Carel de Rooy, "The Decade and Beyond: At a Glance," *Water International*, September 1991; global military spending from Ruth Leger Sivard, *World Military and Social Expenditures 1991* (Washington, D.C.: World Priorities, 1991).

4. Jordan's agricultural water use from Maher F. Abu Taleb et al., "Water Resources Planning and Development in Jordan: Problems, Future Scenarios, Recommendations" (draft), Jordan Country Paper, World Bank, Washington, D.C., June 1991; West Bank aquifer recharge from Thomas Naff, "The Jordan Basin: Political, Economic, and Institutional Issues," prepared for World Bank International Workshop on Comprehensive Water Resources Management Policies, Washington, D.C., June 24–28, 1991.

5. Motor Vehicle Manufacturers Association, *Facts & Figures '90* (Detroit, Mich.: 1990); Gary Allie, American Iron and Steel Institute, Washington, D.C., private communication, April 27, 1992; U.S. Bureau of the Census, *1982 Census of Manufacturers, Water Use in Manufacturing* (Washington, D.C.: U.S. Government Printing Office, 1983, issued March 1986); U.N. Industrial

Development Organization, *Industry and Development, Global Report 1990/91* (Vienna: 1990).
6. Share of grain fed to livestock from U.S. Department of Ariculture, Foreign Agricultural Service, *World Cereals Used for Feed* (unpublished printout) (Washington, D.C.: 1991); Marcia Kreith, "Water Inputs in California Food Production," prepared for Water Education Foundation, Davis, Calif., September 1991.
7. PRB, *1992 World Population Data Sheet*; World Resources Institute, *World Resources 1992–93* (New York: Oxford University Press, 1992).

Index

ABOUT THE AUTHOR

SANDRA POSTEL is Director of the Global Water Policy Project in Amherst, Massachusetts, where her research focuses on international water issues and strategies. From 1983 until 1994, she was with Worldwatch Institute, where for six years she served as Vice President for Research. She has published widely in scholarly and popular publications; lectured at Stanford, Harvard, Duke, MIT, and Yale universities; and, for two years, taught at Tufts University as Adjunct Professor of International Environmental Policy. She has served on the Board of Directors of the International Water Resources Association and the World Future Society, as an advisor to the Global 2000 program founded by President Jimmy Carter, and as a consultant to the World Bank and United Nations Development Programme. In 1995, she was awarded a Pew Fellowship in Conservation and the Environment.

Last Oasis was selected by *Choice*, a magazine for educators, as one of the outstanding academic books of 1993.